T0093649

THE
HIDDEN KINGDOM
OF FUNGI

Foreword by Rob Dunn

Keith Seifert

THE
HIDDEN
KINGDOM
OF
FUNGI

Exploring the Microscopic World
in Our Forests, Homes, and Bodies

GREYSTONE BOOKS
Vancouver/Berkeley/London

Greystone Books Ltd.
greystonebooks.com

Cataloguing data available from Library and Archives Canada
ISBN 978-1-77164-662-8 (cloth)
ISBN 978-1-77100-663-5 (epub)

Developmental editing by Linda Pruessen
Substantive editing by Lucy Kenward
Copy editing by Dawn Loewen
Proofreading by Jennifer Stewart
Indexing by Stephen Ullstrom
Jacket and text design by Jessica Sullivan
Jacket illustrations by Florilegius and Antiqua Print Gallery/Alamy Stock Photo

Printed and bound in Canada on FSC® certified
paper at Friesens. The FSC® label means that materials
used for the product have been responsibly sourced.

The writing of this book was supported by a grant to the author
from the Alfred P. Sloan Foundation's Public Understanding
of Science, Technology and Economics program.

Greystone Books thanks the Canada Council for the Arts, the
British Columbia Arts Council, the Province of British Columbia
through the Book Publishing Tax Credit, and the Government
of Canada for supporting our publishing activities.

MIX
Paper from
responsible sources
FSC
www.fsc.org   FSC® C016245

Canada

BRITISH COLUMBIA

BRITISH COLUMBIA
ARTS COUNCIL
An agency of the Province of British Columbia

Canada Council   Conseil des arts
for the Arts   du Canada

Greystone Books gratefully acknowledges the xʷməθkʷəy̓əm (Musqueam),
Sḵwx̱wú7mesh (Squamish), and səlílwətaʔɬ (Tsleil-Waututh) peoples on
whose land our Vancouver head office is located.

_____

We know that we come from the winds,
and that we shall return to them; that
all life is perhaps a knot, a tangle, a
blemish in the eternal smoothness.

E. M. FORSTER
A Room With a View

CONTENTS

FOREWORD

MYCOLOGY IS A DARK ART. Mycologists perceive things that others miss. They traffic in the scarcely visible. They make meaning out of what they see through hand lenses or what they smell when they turn over logs, tip over mushrooms, or sink their hands into the soil. In amongst toadstools, moulds, and yeasts, they find answers to questions but also evidence of mysteries. Augury is the practice of interpreting signs and omens from the behavior of birds. Mycology can be the practice of interpreting the meaning of the world from the behavior of fungi.

The offices of mycologists are rarely new and gleaming. Instead, they tend to be in basements and other forgotten corners of buildings. Like fungi, mycologists often dwell in neglected places. In those places, any extravagances that might be encountered are typically engineered not on behalf of the mycologists themselves, but instead for their quarry, fungi. At my university, for example, air conditioners poke out of broken basement windows where the mycologists work. The air conditioners keep collections of fungi from getting too hot or too wet; the offices of the mycologists, meanwhile, are not nearly so climate controlled.

In their neglected places, mycologists guard a secret. The world, they know, is not as it seems. If you walk outside, you may hear birds singing and you will see plants busily turning light into sugar. The birds and the plants then suggest the rest of life—mice and insects, for example. But this view of life is wrong, entirely wrong. Beneath the world you see and hear, and inside that world, and over and above and around that world, are fungi. Fungi are busy, nearly everywhere (and perhaps everywhere) there is terrestrial life (though also in the sea), carrying out strange slow-motion sexual dances, eating what seems inedible, signaling to one another and to other species too, and then also connecting organisms, one to another. By most accounting, there are more kinds of fungi than there are mammals, birds, and plants combined, perhaps ten times as many. Fungi are the connective tissue of life—not just that, but certainly that. They hold the living world together. This is what mycologists know. And they see the evidence everywhere they go. It is unavoidable to them, the fungal world of particulars and grandeur. They live in it, they sniff it out, they breathe it in, and, when they get together, they talk about it constantly, using names of species that sound like spells or incantations. Their minds are bubbling flasks of *Saccharomyces cerevisiae*, *Aspergillus*, *Candida*, and *Penicillium*. And maybe even some *Entomophthora muscae*—spore of fly—to boot.

Mostly, the secret world mycologists know, the true world, is secret because it is not obvious and, after years of trying to convince other people to love fungi, mycologists tend to realize it is easier to just talk to other mycologists about what they know and to leave the rest of us out. They retreat to their lairs where they talk about beetles whose brains are taken over by zombie fungi, ants that farm fungi the way that humans grow grain, fungal pathogens that threaten global food supplies, or, just as often, lovely, delicate, poorly known species that they have somewhere seen and just what those species might mean about the broader world.

I have been lucky enough, every so often, to be welcomed by mycologists into their world and, while there, to see things a little more as they are, which is to say to see them a little more fungally. It is a great privilege. Like an astronaut having gone to the moon and then returning to Earth, I return from each visit to the mycologists changed, with a different perspective on everything around me.

It is the extraordinarily good fortune of readers everywhere that Keith Seifert has decided to serve as a guide to the world of fungi, the world as seen by mycologists. In this wonderful book, *The Hidden Kingdom of Fungi: Exploring the Microscopic World in Our Forests, Homes, and Bodies*, Seifert introduces the reader to that which is obvious to mycologists but hidden to nearly everyone else. His is an insider's tour to, as he puts it, the hidden kingdom.

This book provides snapshots of what are, in essence, the great wonders of the fungal world. There are stories of fungal evolution, tree roots and fungi, plant evolution, beer, agriculture, homes, human bodies, and more. In these stories, one confronts the extraordinary diversity of ways that fungi alter and, really, control the rest of life and even some of nonlife. One also finds beautiful details. Seifert has spent his life among the fungi; he knows as much about fungi as anyone knows about anything. Here, though, rather than use his detailed knowledge of the fungal world to catalog each bit of what he knows, Seifert uses that knowledge to heighten the reader's understanding of the big stories; one has the feeling of having been invited as a novice to the mycologists' table. The experience is, at once, surprising, thrilling, and engrossing.

And that we are invited is important, because the truth is that mycologists can lead us to see the world as it is (or closer, anyway, to how it is). They can lead us to understand our own place in things. Such an understanding is important, because as Seifert shows in the final chapters of the book, fungi not only control much of the world, they also offer many of the most extraordinary solutions to our future challenges. The fungi have shown the mycologists the way to a great many answers, and Seifert concludes by holding those answers up for the rest of us. "Look to the fungi. Learn from their ways."

It's become common to describe our geological epoch as the Anthropocene, the human epoch. The Anthropocene is defined primarily as a function of the extent to which the impact of humanity is clear on the composition of life on Earth and biogeochemical cycles of life and nonlife on Earth. It is true that our impact is great and the naming of our time as the Anthropocene reminds us of that horrible greatness, reminds us of the consequences of each of our days and each of our actions. And yet, from the broader perspective of Seifert and the other mycologists, it is clear that we are not living in the Anthropocene. Instead, we are just in one particularly unusual period of a broader epoch, what one might call the Mycocene. Ours is a fungal world, and by the time I finished this lovely book, I was convinced that however great our human impact might be, it pales in comparison to that of fungi, as it long will.

When humans go extinct, our crops will disappear and so too many of the strange species that depend upon us, from German cockroaches to pigeons. Forests and grasslands will grow back. Populations of medium-sized fish and terrestrial and marine predators will rebound. As our presence is felt, so too will be our absence. But, as Seifert shows through his stories of the fungal world, were fungi to go extinct, the situation would be something else entirely. Wood would cease to decay. Tree roots would be unable to get nutrients. Millions of species would undergo population explosions, their abundances no longer kept in check by their fungal pathogens. The climate would be altered in ways so dramatic as to be difficult to even think about, much less predict. The world is so very deeply and comprehensively fungal that to think about a world without fungi is nearly impossible.

And yet, we do think about the world without fungi. We do it all the time. We pretend they aren't there at all. So here is my suggestion. Read Seifert's book. Then, once you have, tell other people about it. Tell them about the fungi. Help them to understand a little bit more about just how wrong so many of our daily perceptions are. Initiate them into the secrets of mycology. Then go outside and take a deep breath. Sink your fingers into the soil. Crawl up upon a mushroom. Swirl and sniff your beer. Look at the discolorations on grains. Practice reading the secret signs. Mycology is a science in which we can all engage, one that begins with something like a little meditation. Go ahead and say it out loud: "I am human in a fungal world, a new arrival in the hidden kingdom." And after you do, breathe in deeply. Let the thousands of fungal species in the air land on your tongue and travel into your lungs. Then exhale. You are surrounded. You have always been surrounded. Mycology is the dark art of knowing this reality; mycology is the dark art of knowing the truth.

ROB DUNN, author of *Never Home Alone*

A NOTE
ABOUT NAMES

ABSORBING THE NAMES OF ORGANISMS can be like trying to follow characters in a Russian novel. Everyone seems to have several names—formal names, family names, nicknames—that are only clear to those steeped in the culture. The rest of us mentally bleep over them and end up confused about who loves whom and who kills whom.

Scientific names are a necessary evil that you can't avoid when talking about fungi, because they are often all we have.[1] Over the past few decades, mycologists have made a determined effort to come up with distinctive common names for conspicuous mushrooms and lichens,[2] but the multitude of yeasts, moulds, rusts, smuts, and mildews were left behind. I use common names in this book when I can because they are easier to remember and pronounce, but I have tried to avoid inventing new ones. Unfortunately, these names don't always translate coherently from English to other languages, and often they don't precisely match how scientists think about the same species. So scientific names are used when needed.

But before you decide to sidestep them and lose track of the plot, let me give you a quick primer. The system for naming, defining, and classifying groups of fungi follows the standard hierarchical structure used in biology, from a single species to gradually larger and larger groups: genus, family, order, class, phylum (*pl.* phyla), and kingdom.[3] Each group has its own latinized label that is usually only used in discussions or presentations of classification.

Everyday language has terms roughly equivalent to a genus name for groups we recognize as sharing common traits—ducks, roses, pines—although these folk categories often do not correspond exactly with how scientists group the same species. There are a few such words for fungi in English; for example, larger mushrooms like chanterelles (= *Cantharellus*) or honey mushrooms (= *Armillaria*).

The two-part Latin scientific name for a particular species is known as a binomial. The first word in the binomial indicates the genus (*pl.* genera). The second word (the epithet) labels the species within its genus. The combined binomial is unique for each species. For example, the scientific name for brewer's yeast is *Saccharomyces cerevisiae*, just as we are *Homo sapiens*. The binomial is set off in italics. An increasingly common practice in mycology is to italicize formal scientific names at all other levels as well (phylum, class, etc.). I have followed this new practice in the appendix (page 221).

I've spent a lot of my career obsessing over Latin names of fungi, and I assure you they can be humorous.[4] Sometimes the jokes are obscure, but who could miss the point of a mushroom called *Spongiforma squarepantsii*? It really does resemble SpongeBob (sort of)! And because no one ever noticed, I'd like to point out that I once named a fungus after Homer Simpson, although my hidden agenda to honor the whole Simpson family in this way never materialized.

Though it may not seem like it, most binomials do tell you something about a fungus—and catching the innuendos makes them easier to remember. Take *Saccharomyces cerevisiae*. *Saccharo* means "sugar" and *myces* means "fungus." If you've ever ordered beer in a Spanish-speaking country, you'll know that *cerveza* is the Spanish word for "beer." So the Latin name describes brewer's yeast perfectly as "the sugar fungus from beer." After all, one reason we love this fungus is that it takes the sugar in barley or grapes and converts it into ethanol.

Some university professors quiz their students about the hidden nuances of scientific names; it's a rite of passage into a broad palette of knowledge. Most of us don't study Latin in school anymore, so parsing out the hidden messages in the names of fungi takes some study. Saying them out loud helps you remember.

Learning Latin names opens the door to appreciating the awe-inspiring diversity of the millions of species of animals, plants, fungi, protists (mostly one-celled organisms, like amoebae), and bacteria that await you in nature. The appendix puts the fungi mentioned in the following chapters into the modern classification system. Think of it as a phone book for looking up the evolutionary address of a fungus.

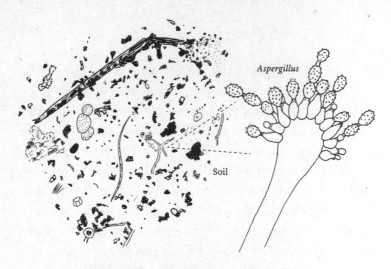

Aspergillus

Soil

INTRODUCTION

Diversity in the Dust

FOR MOST OF US, dust is dust.[1] We don't think much about what it is or what it might mean. It is simply the powder of our world. It drifts onto the floors of our homes and hospitals, blows through farms and forests, and settles on the sea floor. It swirls around the globe, wafting across oceans from one continent to another, from one country to another. Its parts are too tiny for our senses to register; it seems too banal to be important.

When we were kids, mud caked between our toes and in the creases behind our knees as we raced around the yard. We built castles in the sandbox. Our mothers scrubbed our faces too hard with the washcloth trying to make us presentable, to create the illusion that we were clean. But we were always going to get dirty again. It was inevitable.

When you're a kid exploring your world, everything is wonderful and just as it should be. You learn the important laws of life, like "Don't eat dirt." If candy falls on the floor, the five-second rule comes into play: How long can it stay there before it's too nasty to eat? A little dirt, a little dust—if you don't see it, or if you can rinse or wipe it away, it doesn't matter.

But what happens if you take a tablespoon of dust or dirt and stir it into a quart of water? Add a tablespoon of that slurry into another bottle, then press repeat and dilute it one more time. If you looked at the murky blend through a microscope, you'd start to appreciate the complexity of dust. Tiny crystals and mineral chunks mix with flakes of rotten wood, legs and hairs of insects, soot, odd-looking eggs, and fibers from plants and your clothes. Microscopic algae and protists bump into each other and veer off like windup toys. Dust is alive.

If you treat the diluted mud with a dye that binds to DNA—the chemical compound that makes up the genes of all living things—and shine ultraviolet light through it, microscopic life lights up like the Milky Way, a universe in a drop of water.[2] Bacteria and viruses shine like stars. Pollen grains drift by like glowing blimps. And among all these particles are the extending tubular cells, geometrical spores, and budding yeast cells of fungi.

This book is a journey through the hidden world of fungi and their relationships with humans, other living things, and our environment. We will look at how we use fungi, and how they use us, as we strive for a sustainable future.

Looking back, I can see that my family history and childhood experiences set me on a path with an unexpected result. I didn't plan to be a fungus guy—who does?

The Promised Land

My dad's parents emigrated from Germany to Canada more than a hundred years ago. They were patented a quarter section of land in the province of Saskatchewan, purchased three more, then farmed wheat and raised their family through the Great Depression. My parents met in Regina at Normal School—teachers college. When World War II started, my father enlisted in the Royal Canadian Air Force. He was color-blind and couldn't be a pilot or serve overseas, so he worked as a mechanic. His oldest brother inherited the farm. So after the war, with support from the family and the Veterans Rehabilitation Act, Dad's best-laid plan was to study architecture at the University of Manitoba in Winnipeg. When it was time to look for a job, he climbed aboard an eastbound train with my mother and my two older sisters. They disembarked and settled in the mining town of Sudbury, Ontario, where my third sister and I were born. Perhaps the desolate wasteland reminded them of rolling prairies.

In the late 1800s in Sudbury, metals were mined in giant open pits on the edge of town. Stacks of layered rocks and wood were set alight and smoldered day after day. Dense sulfur smog rolled over the lips of the pits and across the hills, killing plants, blackening granite, and leaving treeless barrens that lingered for generations. Industrial smelters replaced the so-called roast yards, but the chimneys only carried the soot so far. The city resembled a grayscale pencil sketch because of the 116 tons of nickel, copper, zinc, and iron that belched from the smelters every day.

Starting in 1970, the Inco Superstack steadily rose above the city's western horizon. After two years of construction, the completed 1,250-foot chimney discharged clouds of sulfur dioxide and nitrogen dioxide into the upper atmosphere. The plumes showed the direction of the prevailing winds that carried the noxious gases and acid rain to distant northern Europe. Sudbury's desolate lands had drawn attention from NASA by then. The basin that cradles the city is the remnant of a huge prehistoric meteorite crash. In 1971 and 1972, before their missions, the Apollo 16 and 17 astronauts tested the lunar rover outside town and studied the geology of an impact crater similar to what they would see on the moon. This degraded environment was my home. It wasn't a typical introduction to the love of nature, but to me it was still a world of wonders. The thrill of space exploration led to my love for science.

My parents planted rock gardens with vegetables, fruits, and a wildflower or two that tolerated the city's polluted air and acidic soil. My sisters loved these plants and the scruffy weeds sprouting in the gravel of vacant lots. On summer weekends, my father drove for hours—my sisters in the back, me propped between my parents in the front—along the winding, roller-coaster roads outside Sudbury. If we got far enough from the mines, there started to be tree cover. Long stretches of road were punctuated by Do Not Trespass notices. If Dad found a stretch with no signs in sight, he led us over the fence. He told us that the barrier was not for us, but to prevent bears or moose from wandering onto the road.

Our favorite spot was within sight of the white, quartz-laden hills near the town of Killarney. We scrambled through gullies between granite hills scraped and scratched by ice-age glaciation to reach the shore of the Bay of Islands on Lake Huron. Extended patches of shoreline, in between the acrobatic jack pines, were blemished with black, leathery lichens. If it rained, they swelled into slippery, rubbery scabs. Any misstep ripped the root-like anchors of the lichen off the rock and sent me skidding onto my knees. "It's called rock tripe," Dad said—one of the few snippets of mycological trivia he ever shared with me. "Some people say you can eat it. Supposedly it tastes like scrambled eggs." And then, relaying something from one of the history books he was always reading: "They made soup out of it on the Franklin Expedition."[3]

I never met anyone who could comment on its flavor, but I did learn its scientific name: *Umbilicaria*.

These expeditions weren't quite roughing it, but they were still agony for me. My sisters stopped to discuss every plant. They picked edible flowers for salads and boiled the pith from the inside of swamp cattails and served them up like exotic vegetables. I was more interested in the skinks and worms that our dachshund unearthed in her frantic excavations of soil and rotting wood. Her short legs whirled like rototiller blades when she dug holes. Surfacing, she'd glance over with a demented, conspiratorial grin, encouraging me to share her discoveries, thrusting her snout deep between the roots. The smells! The smells! She perceived things that I missed. I wanted to smell the world the same way that she did. Taking her as my role model, I started to appreciate details in the surroundings that I'd never noticed before.

With little sense of direction, I entered university as a science student. After false starts in astronomy and biochemistry, I emerged nine years and three institutions later as a specialist in mycology.[4] My fellow grad students and I had the mad idea that "mycologist" was a prestigious job. Mycology is a profession where you spend a lot of time explaining what it is you actually do, and when people finally understand that you study fungi, they usually don't believe you. Who would pay you to play around all day with moulds and yeasts and mushrooms?

Despite the absence of a sensible career path, as the years ticked by I remained a contentedly employed researcher studying farms, forests, and the built environment, spending time on five continents along the way. Sometimes I stopped crawling in the duff and gazed up from the diseased plants and rotting logs long enough to notice a few prosaic tourist attractions—but just sometimes.

The Fungal Kingdom

Most people are unaware of fungi, although we pass them every day and inhale their spores with every breath. Fungi are stereotyped as agents of decay, disease, rot, and mould, spoiling everything that is clean and pristine. We tolerate moulds in compost buckets but not on our bread. We have strong opinions about whether mushrooms are acceptable as food. The rest—the thousands of species we encounter every day— remain unseen and unimagined.

We understand the larger, conspicuous fungi—or macrofungi—best.[5] Mushrooms are the most familiar, but they are transient structures that last just a few days. Our preoccupation with food makes us wonder how to distinguish poisonous and nonpoisonous mushrooms—a recurring plot device in murder mysteries.[6] Edible fungi, like oyster mushrooms, shiitake (it's important to remember the double i), porcini, chanterelles, morels, and the underground truffles, keep naturalists and chefs busy for a lifetime. Even lichens (see chapter 2) are sometimes used as food, or to dye clothes.

This book focuses on the microscopic fungi that we rarely notice and understand so poorly. They are commonly called moulds, a casual term that covers thousands of distantly related fungi, just as the word "shrub" is used for unrelated plants sharing a similar pattern of growth. Moulds are usually betrayed only by a dusty, cottony, slimy, or powdery haze, sometimes surrounded by a faint array of filaments. Most fungi, including macrofungi, spend the majority of their lives as an almost invisible network of microscopic threads (*hyphae*). Occasionally, some form larger structures (for macrofungi, again, these may be "mushrooms"), which release clouds of nearly invisible spores. The spores float through the air and settle almost everywhere, including on our food and in our beds.

Over our long history together, fungi have often been our rivals, but they also help us out quite a lot. We often join forces with single-celled fungi called yeasts. There are thousands of wild species, but a few are essential for producing our staple foods and drink. Yeasts grow in many sorts of liquid, including the water-saturated bodies of insects and humans. There they help keep the digestive tract ticking along as part of our friendly gut flora, and they form part of the microbial coating that protects our skin.

Moulds are critically important hidden partners on farms and in forests as intimate associates inside plants and animals. We adapt many chemicals fungi make for their own purposes as medicines like antibiotics. Fungal enzymes—proteins that break down, put together, or rearrange other molecules in biochemical reactions—are used as additives in industry to boost detergents or to help make biofuels. They were among the most successful early products of modern biotechnology.

Unfortunately, when we aren't looking, fungi also cause problems, such as plant diseases like rusts, blights, smuts, mildews, and cankers. With their talents for biodegradation—breaking down organic matter—fungi rot out the floorboards of our houses, or spoil our food and lace it with toxins. Doctors are familiar with itchy fungal skin conditions, like dandruff, ringworm, and athlete's foot, and more frightening infections called mycoses. And like some human viruses that concern us, fungi sometimes jump from continent to continent causing new diseases to spring up in distant locations.

Despite the fact that humans and fungi have different body designs, our cells and biochemistry have a lot in common.

This similarity makes fungi useful for medical research, but it also means that if we try to hinder them, we must ensure our chosen weapon doesn't rebound on us. Chemicals toxic to a disease-causing fungus might also affect humans. This is one of the reasons we have so few effective antifungal drugs, and why fungicides for crops should be assessed so carefully.

Cultural attitudes towards fungi vary from one society to the next, depending on how people weigh their helpful and harmful properties. So many Westerners have such an inbred revulsion towards fungi that there is an adjective for it: *myco-phobic.* In many parts of the Western world, "fungus" and "mould" are punch lines for jokes and used as intentional insults or indicators of moral degeneracy or poor hygiene. But many northern or eastern European, Asian, and Indigenous cultures regard fungi with an affection similar to what we usually reserve for kittens or puppies. The adjective then is *mycophilic.*[7] As an example, one of the most popular anime characters on Japanese television is the kōji mould (*Aspergillus oryzae*): a happy round yellow face, with five small cones of stacked round spores radiating outward, who bobs through the air smiling and singing.

Why this difference in cultural attitudes? It reflects the differing responses, fear or curiosity, that we feel when confronted by the unseen parts of nature. Fungi can be good or bad, but most are somewhere in between. Now, with so much effort in society to correct ancient wrongs, the time seems right to set aside our prejudices.

The Fungal *Umwelt*

The German word *Umwelt*, a poetic and philosophical convention, imagines how animals perceive their surroundings with their eyes, ears, and brains. We are confident in the reliability and completeness of our animal senses. We seldom imagine that other living things have capacities we lack or that they send and receive different kinds of signals than we detect. Our lack of empathy—our impaired *Umwelt*—is reflected in how we treat other living things. We enjoy a modest camaraderie with animals of a certain size: our pets and the charismatic mammals on television or in zoos. Their babies remind us of our own babies. They relate to us in some social manner. But our warm, fuzzy feelings dissipate when we consider other animals. How do you feel about insects? Or frogs or bats? Peering into the eyes of these creatures feels like staring into the eyes of an alien. Empathy for invisible life-forms seems a long way off.

Humans are often surprised by the complex behavior of microorganisms. It is easy to dismiss microbes—microorganisms like fungi, bacteria, slime moulds, and protists—as miniature machines, or automata, that act and react with a predictable mechanical response to external cues. This philosophical approach is called animal chauvinism or speciesism. It's a barrier to recognizing decision-making, creativity, and any sense of control or agency in other forms of life. At the very least, fungi are living beings that react, eat, excrete, send out and receive signals, mate, and strive for a better life. In those ways, they are just like us.

Interpreting the behavior of other living beings as equivalent to human cognition, emotion, or agency is anthropomorphism. In science, this is often considered an unforgivable error. But the very notion of anthropomorphism seems anthropomorphic. We too are locked into our version of existence by our senses and consciousness. Anthropomorphism is the best tool we have to imagine creatures that are so dissimilar—that operate at such diverse scales, move around by such peculiar means and at variable speeds, and transmit and receive different signals than we use to communicate ourselves.[8]

Imagining our world from the point of view of a fungus is a challenge, but because this book is about fungi I will be unapologetically fungopomorphic (or, if you prefer, mycopomorphic). But an analogy is just an analogy. It's a tool to help us understand and empathize. I am not a fanatic (a fungatic?). I don't pretend that fungi are more important or interesting than the other kingdoms of life. But I am an unashamed fungal partisan. We are living in an era of declining biological diversity at the same time as we are becoming aware of an unexpectedly vast interconnectivity among all life. In this book, fungi are both the heroes and the villains—humans are just the supporting cast.

Fungi are our close neighbors in the evolutionary tree of life, more closely related to animals than to plants. We think there may be between 1.5 and 15 million fungal species (with 5.1 million a reasonable compromise), but only about 140,000 are cataloged and named despite two hundred years of study by mycologists with microscopes. This means we may have seen and classified less than 5 percent of them.[9] Over the past twenty years, DNA signals have revealed the presence of an unexpectedly large number of unknown species. Slowly, we are filling in the missing pieces of a greater puzzle.

I hope this journey through our neighbor's kingdom will help you appreciate the complexity of the living world and the need to acknowledge, understand, and respect all organisms, no matter how small.

Let me introduce you to some of my friends.

PART 1

THE HIDDEN KINGDOM

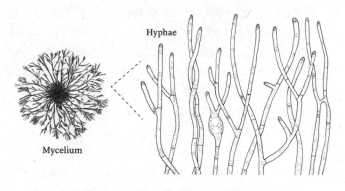

Hyphae

Mycelium

1 | LIFE IN
THE COLONIES
Fungal Evolution

EARTH DID NOT ALWAYS LOOK like it does today. The land
we stand on feels solid, but the continents float on molten
magma like dumplings on a simmering stew. For the first billion years or so, most action on the planet was geological and
chemical. The ancient, lifeless supercontinents drifted apart
and bashed back together, most recently only a quarter billion
years ago as a single landmass called Pangaea.

Life appeared about four billion years ago during the geological period we call the Precambrian Era. For the first two
billion years, sometimes disparaged as the Boring Billions,
all life-forms had just one cell. The atmosphere was a blend
of nitrogen and carbon dioxide gases. Life is sweet: it's all
about sugar. A few bacteria discovered how to use sunlight
to make sugar from the air by stitching carbon dioxide and
water molecules together. This biochemical reaction, called

photosynthesis, provides the food for all life. Oxygen was a by-product. As the millennia passed, the concentration of oxygen in the atmosphere climbed to 21 percent. Other microbes scrounged the waste secretions of these photosynthetic cells, survived on their tiny carcasses, or attacked living cells directly—life surviving off death by the processes of saprobic or parasitic nutrition.[1] Between 1 and 1.5 billion years ago, microbes with nuclei and multiple X-shaped chromosomes (known as eukaryotes) split into various kingdoms of life—including animals, fungi, plants, and several groups of protists.[2] The bacteria (often called prokaryotes), which have no nuclei and usually just one circular chromosome, went off on their own trajectory. Based on sheer numbers of cells and species, bacteria still dominate the modern world—but that's a different story.

We think that the last common ancestors of animals and fungi were single cells that swam in the ocean with a whiplike flagellum (*pl.* flagella): microscopic tadpoles called *zoospores*. The ancestors of fungi were so tiny and fragile that only a few fossils exist that provide clues to what they looked like. Today, the surviving offshoots of this ancient evolution belong to the phylum Chytridiomycota, known as chytrids. The majority of the thousand or so species live a bucolic life in freshwater. Their zoospores wag their tails and flutter from place to place. They ram their way into pollen grains or the epidermis of floating seeds, like tiny goats butting a balloon. Then they swell up into one or a few cells that fill up with zoospores again. Some make finger-like cellular roots called rhizoids that grab on to or penetrate the host tissue. Sometimes zoospores smash their heads together and mate. You have probably never heard of chytrids (the name comes from the

Greek for "little pot" and describes the mother cells filled with zoospores), but their affection for moisture and their habit of setting up zoospore factories in plants and animals lead to some serious diseases. The well-known amphibian apocalypse (see chapter 9) is one example.

Zygomycetes ("zygos" for short) were the next fungal group to split off, when modern multicellular life began to diversify and move onto land. Most modern zygos (now reclassified into several phyla; see appendix) are fast, weedy moulds like the compost mould *Rhizopus*. They search out moist, sugary nooks, pop up quickly, and saturate their surroundings with asexual spores. Quite a few are involved with insects. The familiar *Entomophthora muscae* glues houseflies to your windows each autumn and spatters a halo of white asexual spores around the corpse. The sexual process in zygomycetes is more like ours than most other fungi because there is only one child per mating. Pregnancy happens outside the body, though—the sexual zygospores swell up between the tips of a mating pair (see figure on p. 223). These dark, thick-walled balls are often covered with warts or elaborate branched projections that under a microscope give them the look of steel wool or the working end of a medieval flail. There is no mechanism for sending these spores elsewhere; they just drop into the dust and hibernate until favorable conditions return. Their dramatic ornamentation protects the spores from being devoured by hungry insect larvae and nematodes.

Root-associated zygos called arbuscular mycorrhizal fungi (see chapter 4) are among the fungi that helped plants adapt to life on land. Those cartoons of a fish squiggling out of the sea, growing legs, wobbling along, and transforming into a lizard, then a dinosaur, then some kind of ape, and finally, after millions of years, Homer Simpson—it didn't happen like that. Although the details are lost to time, most land plants probably evolved from multicelled algae that migrated into freshwater, away from the salty violence of the sea. In that first few hundred million years, algae and fungi bobbed together in shallow ponds or dried together into crusts. The intimate cooperation and competition that still exists between the fungal and plant kingdoms bloomed.

Plants wouldn't be as successful if they weren't involved with fungi. Many cooperative relationships between fungi and plants developed and continued through the eons, eras, periods, epochs, and ages into the modern world. Hyphae—the thread-like cells that make up the bodies of fungi—grew inside plant leaves, stems, or roots. A few plant fossils have been found that give clues to these ancient relationships. Another strategy was for single-celled algae (or photosynthetic bacteria) to live inside colonies of a fungus, an arrangement we now call lichens.

The proliferation of fungi and land plants was dramatic and seemed to happen in sync. About 400 million years ago in the Devonian Period, the two largest fungal phyla of the modern world, Ascomycota (the ascomycetes) and Basidiomycota (the basidiomycetes), appeared. Both interact intimately with living and dead plants, but in different ways.

Ascomycetes, or "ascos" for short, are the largest fungal phylum. You may know some of the conspicuous ones. Lumpy, coal-like, pricey truffles (*Tuber* species) nestle among the roots of oak trees. If dogs, wild pigs, or squirrels don't find them first, you can dig them up and shave off some slivers to enhance your finest meals. Elusive wrinkled, egg-shaped morels (*Morchella* species, sometimes called Gucchi mushrooms) pop up in meadows and under trees for a few weeks each spring. They are prized by mushroom foragers, who often exhibit strong hoarding behaviors and keep their precious locations a secret.

Many of the roughly 87,000 known ascos are microscopic. You see them only as little dots or blobs on plants or animals. Or they hide away in decaying debris nibbling at the cellulose or starch, then bloom as an asexual, spore-spewing mould. Their sexual states are flask- or cup-shaped bodies that emerge from a nest of intermingled parental hyphae. Sexual spores form inside a larger sac-like cell called an *ascus* (*pl.* asci) that bulges out from the cell where the nuclei of the two parents merge. If the nucleus inside each ascus divides once, the fungus ends up with two ascospores per union; if twice, then 4, then 8, 16, 32, sometimes hundreds. The typical ascus, though, has eight neatly arranged ascospores and looks something like a transparent pod or sac full of beans. In most species, the asci are squirt guns: they rupture and a gush of cellular fluid propels the spores out into the air. Hundreds or thousands of asci may ripen at once, each the result of a unique union of nuclei from the original parent hyphae. It's quite the orgy.

The ability to break down both cellulose and lignin in plants is a distinctive talent of many species of the other large fungal phylum, the basidiomycetes ("basidios"). About half of the fifty thousand species form what we generally consider mushrooms. Some have caps with gills, like the typical brown grocery store mushroom, *Agaricus bisporus*. Others have a spongy underside covered with tiny pores—boletes like the cep or porcini (*Boletus edulis*) adored by European mycophiles. Polypores (like the reishi mushroom, *Ganoderma lucidum*) are tough, woody bracket fungi that decay trees and lumber and have thousands of speck-like pores on the underside of their shelf-like structures. Puffballs are round fungal marshmallows about the size of a golf ball that soften with age and dispatch puffs of grayish spores into the air. (The giant puffball, *Calvatia gigantea*, swells to the size of a soccer ball and often gets treated like one.) Jelly fungi, like the yellow witch's butter, *Tremella mesenterica*, often found on tree branches, swell into brain-like gelatinous masses when wet but shrink to a hard scab when dry. Some have yeasts as asexual states. Many microscopic basidiomycetes lurk in soil or plants too, notably the pathogenic rusts and smuts that can be so devastating to agriculture.

In many basidios, the hyphae arrange themselves into a miniature, concentrated mushroom in the soil and wait. When the showers come, the cells of this primordium inflate with liquid and a complete mushroom bursts out of the ground like a compressed sponge dropped into water. The force of this swelling can even break concrete. The gill or pore surfaces of the mature structure are covered by a layer of thousands of club-shaped microscopic cells called *basidia* where the nuclei originating with each parent merge. Each basidium is crowned with four tapering pins known as *sterigmata*. Each sterigma supports a developing basidiospore balanced off-kilter on its perch, and these spores inflate in synchrony to their final round or oval shape. Using a mechanism that involves the explosion of a tiny droplet of water, the spores pop off into the space between the gills and fall into the air current streaming around the cap.

Contemporary mycologists count about twenty phyla in the fungal kingdom.[3] About sixteen of these groups contain only a few species and have little impact on humans. But the A, B, C, Zs—the ascomycetes, basidiomycetes, chytrids, and zygomycetes that we just met—together encompass millions of species. They display a vast range of beneficial and harmful behaviors that influence both nature and human civilization. To see them, you need only learn how to look.

Learning to See: Bringing Fungi Into the Light

How might you describe yourself on the telephone to the stranger who volunteers to meet you at the airport? I would be one of many slightly chubby, blue-eyed, shortish, middle-aged men with straight brown hair. You might need some special vocabulary to describe the shape of my nose, the mole above my lip, and the way I walk. It would be easier if we all wore name tags or had some kind of barcode.

For me, the process of learning to see the hidden world of fungi and distinguish its members started the summer after my second year at the University of Waterloo. In our survey course on nonvascular plants, I'd discovered an unexpected attraction to mycology and wanted to test-fly my new "expertise" on home turf. So at the end of the summer before third year, I stepped into the scraggly woodlot behind our house in Sudbury. I picked a few mushrooms but quickly realized I didn't know much about them. The local bookstore had a single copy of just one guidebook, *The Mushroom Hunter's Field Guide*, by the doyen of American mycology, Alexander H. Smith (1904–1986).[4] To identify mushrooms, I discovered in that book, you make a spore print by resting a cap gill-side down on a piece of white or black paper and cover it overnight with a jar. Or better, half on white paper and half on black, so you can easily distinguish prints of white or black spores. The spores fall off the cap, and in the morning you can see what color they are.

After hours of navigating backwards and forwards in identification keys, and comparing my specimens with the bewildering jargon in the descriptions, I was 70 percent certain that what I had was the honey mushroom, *Armillaria mellea.* As the common and Latin names indicate, the caps are vaguely honey colored and have scales that look something like the crystals that form in old honey. According to the guide, an important distinguishing feature of honeys is the *rhizomorph*, a black shoestring-like strand found around the base. I tromped back into the ravine and scratched through the dirt with an intensity that gratified the dog, who was obviously pleased that I had learned something from her after all. Now that I knew what to look for, the rhizomorphs were everywhere, beneath the loose bark of rotting trees and meandering through the soil and leaf litter like long strips of black licorice. This made me more confident that I had the identity right. Honey mushrooms are also what Smith called "edible and choice," a more enticing invitation than "boring but won't kill you." My goal was to show up my botanical sisters by bringing a wild mushroom to the table.

One rule of wild mushroom consumption is that the first time you eat any new species, you eat just a little. So I told my mother my plan and left most of the mushrooms untouched and the guide open at the appropriate page in case the poison control center needed them later. I heated up the skillet with a pat of butter and fried up one or two mushrooms. How bad could they be? They tasted okay, a bit like regular button mushrooms but with a slight metallic tang. I sat back to wait. Nothing unfortunate happened. Three weeks later, I was in the clear.

Mushroom guides are not shy about describing the dangers of mushroom poisoning.[5] Although most mushrooms are not lethal to humans, the poisonous ones are surprisingly common. Different mushroom species make different toxins, and they affect the body in different ways. As soon as two hours after dining on a mushroom that causes gastric upset, but often as long as seven or eight hours, you vomit or scramble off to the toilet. Others take even longer to make you sick. The fatal effects of the kidney toxin orellanin, produced by *Cortinarius orellanus* and related species, are sometimes delayed by two or three weeks.

The beautiful, juicy-looking *Amanita* mushrooms seem to invite sampling, but species of this genus cause the most fatalities. Their common names convey the correct message. The white North American species *Amanita bisporigera* and the European *A. virosa* are known as destroying angels. The greenish to brownish *A. phalloides*, a European species spreading in North America since the 1930s, is called the death cap. So all mushroom pickers are taught to recognize *Amanita* first, before they learn the edible mushrooms in other genera. The *Amanita* toxins, circular peptides like amatoxin or phallotoxin, survive cooking and pass from the stomach into the bloodstream. Twelve to twenty-four hours later the unpleasantness begins, with severe dizziness and headaches, nausea, diarrhea, hyper-peeing, coughing and shortness of breath, and back pain. There may be a brief respite after this first act, which offers a false sense of recovery as the toxins painlessly concentrate in the liver and kidneys. Then protein synthesis stops and cells burst, leading to an unpleasant death a few days later. Although urban myths suggest treating *Amanita* poisoning with vitamin C or penicillin, these remedies don't help much. The only hope is enough intravenous fluids—and short-term support of vital functions—to buy enough time for a transplant from a compatible liver donor.

You can appreciate that identifying mushrooms precisely is of practical importance, especially if you want to eat them. Misidentifications of fungi cause problems in all kinds of situations, from research labs to farms and hospitals. But how do you identify something you can hardly see? You can only get so far with a jeweler's loupe. For moulds, yeasts, and other inconspicuous fungi, this is a technical process that begins with a microscope. Some citizen scientists do learn these skills. But the need for expensive lab equipment and the perceived need for a university degree often draws a boundary between the hobbyist and the academic. Nonetheless, a few of these lab procedures are important to understand.

If we want to do experiments with a fungus, we need a living culture. Capturing a fungus and getting it to grow in the lab is sometimes easy. You pick up a spore or piece of the colony and place it onto an agar medium in a Petri dish. With a bit of luck, you have the right nutrients and the right temperature, and after a few days the hyphae start to spread. The colonies can be fuzzy or slimy, waxy or powdery, furrowed or smooth, with rays or rings of color, and sometimes with hazy pigments leaching ahead of the growth. Some are fruity, many are musty, and others have no smell at all. You can store purified cultures in refrigerators for months and in freezers for decades, then start them growing again and put them to work. Many fungi don't culture easily, though, because they need a living host or some nutrient is missing or they consider agar a questionable imitation of their homes in nature. Some fungi, maybe the majority, may not grow in culture at all. If you want to study bases, an uncontaminated culture is best, although people with appropriate microscopes and steady hands can start with a single spore.

In modern biology, the most convincing evidence is DNA left at the scene of the crime. Much of the identification of microfungi has graduated away from the microscope into DNA sequencing. This is a practical development because there are so many more people with skills in molecular biology. Anytime you see a picture of a molecular biologist, they are holding a pipettor. This is a narrow plastic cone with a plunger and a stack of ring-like dials at the top, which looks like it should be attached to a video game. Its purpose is to suck up and spit out precise but minuscule volumes of liquid—enzymes, salt solutions, or DNA extracts. Molecular labs overflow with racks full of snap-cap plastic tubes, disposable plastic tips, tubes with solutions, and wonderful little vibrating vortex mixers.

The polymerase chain reaction (PCR) is used to duplicate snippets of genes until their concentration is high enough for chemical analysis. DNA sequences of PCR-amplified genes—the bases adenine (A), cytosine (C), thymine (T), and guanine (G), or ACTG, mixed up in all sorts of ways—are determined with a very accurate method called Sanger sequencing. Technicians used to plop tabloid-sized plasticized sheets called sequencing gels onto light boxes, or hold them up against the window. Each base has its own lane filled with a ladder of black bands. To read the sequence, you follow the bands upwards, jumping from ladder to ladder one rung at a time, each step representing one base. Nowadays, Sanger reactions flow through narrow gel-filled tubes past lasers that detect the DNA. The output comes as a graph with different-colored peaks to represent each base, but most of the sequence is determined by software. In about 2010, a suite of new but slightly sloppy methods was introduced, called next-generation sequencing (next-gen).[6] Because it is faster, cheaper, and has much higher throughput, next-gen is used more and more. Some methods don't need PCR-amplified DNA and can read the sequence from strands of single DNA molecules. Next-gen methods are now used for DNA-based surveys of oceans, forests, farms, foods, plants, animals, buildings, and ourselves, to see what microbes live there.

Whatever the DNA sequencing method, the precise DNA sequences of standard genes are used as genetic fingerprints, which are like DNA barcodes for specific species. Detect the barcode and you can identify the species. DNA sequences are also used to generate the evolutionary genealogies called phylogenetic trees, which assist with the classification of species into phyla, genera, and other taxonomic categories. Differences between sequences can be used to calculate dates in evolutionary time when groups separated, a technique called the "molecular clock." While DNA sequencing helps us decide what fungi we are looking at and how they are related to one another, a more interesting biological question is "What do they do?" To answer that, we need to step out of the lab and into nature.

Hyphae, Mycelium, and Spores: A Fungus for a Day

What if Google Earth didn't stop a few thousand feet above your house but just kept zooming in?[7] Put yourself in Alice in Wonderland's shoes and take a bite of the magic mushroom. Let your body shrink until it is ten thousand times smaller than that awkward blob of limbs and organs you occupy every day of your life. You land softly in a landscape that stretches into the distance. What were once microscopic moulds now sway above you like trees, their spores drifting by like balloons on currents of air. Some of the other creatures and plants around you are so tall that they seem a mile high.

How do you find your way in such a curious scene? That would depend on where you were when you fell down the rabbit hole. If you were in a forest, the matrix would extend deep below and far above, with dirt, roots, worms, and insects teeming in the metropolis around you. If you landed on the back of an animal, you'd crawl out from behind its fur and search for a safe place to hide. But for now, just imagine you're at home and that you've rematerialized in the muddled detritus under the kitchen table.

Looping human hairs and animal fur twist into the sky. Flakes of dandruff stick to some of the strands or sag in loose waxy mounds on the floor. Fibers from clothing and furniture droop like tissue paper streamers. Crashed pollen grains the size of airships lie cracked open and leaking on the ground. Mineral boulders and charcoal briquettes of soot block your path and hide the crevices between the floor tiles. Nematodes several times your size wriggle through the detritus and eight-legged mites thunder around like tanks. Spear-like shards of molted insect exoskeletons jut out of piles on the ground. A few creatures and microbes in your vicinity might sniff at you, but most won't bother—don't take it personally. Humans are unimportant here.

To get our bearings in this miniaturized world, let's turn the looking glass around and imagine that you are a fungus.[8] Your body is composed almost entirely of hyphae, cylindrical cells that are like continuously elongating pieces of spaghetti. They are rarely as wide as a human hair and typically about fifty times narrower. The "skin" around these tubular cells is stiffer and tougher than you are used to, but still flexible. Instead of relying on keratin, the fibrous protein that gives structure to animal skin and nails, fungal cell walls are wrapped with a tape of polysaccharide fibers mixed with chitin. (Polysaccharide means "many sugars"; various sugar molecules are bound into branched or unbranched chains.) Chitin is made of long chains of the sugar N-acetyl-glucosamine, the same compound that hardens the shells of insects and shellfish.

Having filaments as your main construction material might seem limiting, but hyphae can poke their way into narrow passageways or needle their way through soft barriers. They can weave together in various patterns like threads of a fabric. Hyphae provide flexibility and strength, whether they unite into a larger tissue like a mushroom or tie other material together. And outside, there are a lot of hyphae around. Estimates vary from one kind of soil to the next and from one place to another, but nearly 2,000 miles of hyphae—roughly the distance from Paris to Cairo—wind through every teaspoon of rich organic soil.[9]

Try growing a few of your hyphae alongside each other and send some branches sideways. The tips of some will bump together with neighboring hyphae and meld into an irregular lattice. This joining of your different threads into a network is a peculiar fungal routine. Every time you encounter another hypha, you need to run a genetic checklist to make sure it is part of you, or a very close relative, and not someone else. It's risky to join forces with a stranger. Cells or discrete colonies that are genetically identical are called clones. The cloning habit of fungi allows them to link together into larger networks and is one reason they are so ecologically successful.[10]

Eventually you end up as a three-dimensional, tangled pot of pasta called *mycelium* (*pl.* mycelia). As a fungus, you will live most of your life in mycelial form, underground or underwater, in rotten wood or organic debris—one reason why fungi are called the hidden kingdom. Being a bit more fluid than bony animals, you can ease yourself down wherever you are and spread in a meshy network. And rather than being restricted by the need to maintain an organized structure like the body of an animal or a plant, your mycelium exists as a colony. This is a free-form pastiche of connected bits and pieces that is as close as many microfungi get to organizing their mycelium.

Moving from place to place by extending a hypha is slow. To make a quicker jump, you push individual hyphae—or larger bundles or tangles of mycelium in structures like mushrooms—above ground to make spores. A lot of fungi make spores that are single round cells, but some stretch into any number of cells and shapes that look like stars, bananas, hats, bells, worms, or spaceships and are decorated with warts, antler-like growths, or gelatinous tails. Whatever their shape and size, spores contain complete sets of genes wound into chromosomes encased in cells, each like a message in a bottle containing instructions to make a new colony.[11] The drifting spores fly off into the sky like a mass of helium balloons. Their concentration in outdoor air varies a lot but is often a hundred or more times greater than the amount of pollen. Eventually they settle, nearby or far away, falling onto the ground or, with a little dab of glue that sweats through their cell wall, sticking to a plant or insect. Spores sometimes rest for years in soil like a dormant seed, burning small sparks of energy and absorbing a few molecules of oxygen, waiting for the

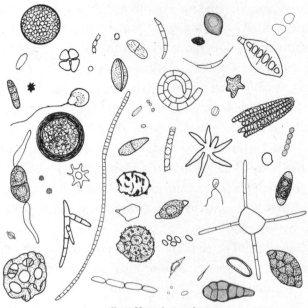

A sampling of fungal spore shapes

right temperature or amount of water, or a special signal. Or they might just go for it and hope for the best. The slumbering genes in the chromosomes crank up to produce enzymes that soften a part of the spore's cell wall. A bump swells out into a new hypha, which extends tentatively into the new surroundings to search out food.

And as a fungus, you will be hungry all the time. Your hyphae can only extend at the end, so to satisfy your constant cravings, you send out more tubes in all directions to search for food. Your hyphal tip is a sensitive probe attracted to water and nutrients. It monitors temperature, gravity, and light as you grow. Typically, it takes a few days or a week to grow an inch, although speed demons like the red bread mould, *Neurospora crassa*, need only six hours.[12] The tip is also where enzymes ooze out to break down plant and animal tissues for food and energy. Your mycelium pushes out in a fan shape, winds into cords, or expands in a doughnut-like ring. The mycelium flows over wood like water, and the resulting pattern morphs into an irregular patchwork. Time-lapse videos of fungal colonies advancing over and around wood blocks show hyphae starting and stopping and changing direction. It looks as if the mycelium pauses to think about which way to turn. The colony sends out and recalls hyphal sentries, evaluating signals—for example, whether there is better food in one direction, or an enemy or a potential mating partner nearby—and then readjusts its growth.[13]

If you are a typical fungus, you will be a *saprobe*. Unable to make your own food as plants do with photosynthesis, and unable to attack a living host filled with easily absorbed sugars, you will wrap your hyphae around some dead organic debris. Biodegradation is the breakdown of organic matter, either waste products from living cells or the dead carcasses themselves, by saprobic microorganisms. For fungi, it is the equivalent of digestion. Since you have no hands or mouth, you pump more than a hundred different enzymes into the surroundings to sop up the minerals, vitamins, and sugars that you need through your hyphal tip for further biochemical processing. Crane-like molecules on the outside of your cells use grappling hooks to grab specific molecules, especially sugars, and pull them into your cells through specialized channels and pores. Breaking down these sugars and polysaccharides gives you energy. After you finish, the leftovers— burps and other gases, all the by-product molecules, ions, and water liberated by your cellular biochemistry—flow back into your surroundings. It may seem gross, but almost all living organisms eat and excrete this way. Animals just hide this process inside.

Day after day, month after month, as a fungus you inch your way through your environment, away from your starting point. If you are blessed with a generous hyphal tip, it dispatches some nutrients back to the older cells left behind, like a grateful son or daughter sending care packages home. But as you continue to grow outwards, the old parts left behind die and slowly dissolve. This releases the precious nitrogen locked up in your proteins and your cytoplasm for reuse by the younger part of your colony. Wherever you are, you run out of nutrients or water eventually. The only thing to do is escape. That's when you should make spores, which often happens in the autumn.

Sex is not your first option. Odds are that you are one of the fungi that make more than one kind of spore. It's far easier to pump out thousands of asexual spores. They will all be identical clones of you, and the softest breeze will send them swarming like shock troops to attack new territory. Asexual spores skip the genetic exchange that defines sex. If you are one of those fungi that can make more than one kind of spore, and need more flexibility than a clone can offer, then sex is the way to go.

Humans tend to focus on the emotional and physical aspects of sex, overlooking the nifty evolutionary benefits that arise when genes from two individuals can mix. The first step of your fungal quest for union is familiar—you have to find a partner of a compatible mating type. In fungi, there are "males" (seekers) and "females" (receivers), which with an atypical scientific display of gender sensitivity and characteristic lack of poetic imagination, mycologists call A and a. Let's say you are an A and want to meet an a. There are hormones for that. Your hyphal tip will aim towards any signal that suggests a receptive partner. When you find one, the a will look pretty much the same as you. Your hyphal tips will examine each other for a moment like two curious dogs booping noses. The same genetic compatibility quiz that you used to recognize your own clone reverses to make sure the cylinder of interest is from a different clone. Your tips fuse and your nuclei dance around inside the conjoined cell, then merge. The genetic gymnastics begin, the swapping of pieces of chromosomes and re-sorting of genes. Rearranging your genes increases the chance that your offspring can cope a bit better in a new environment. With luck, they will inherit some improved enzymes.[14]

If all goes well, a few more cell divisions and you will have four or eight spores (or sometimes many more). Unlike more nurturing humans, your uncaring ascus or basidium will blithely toss your spore children out into the world with only the nutrients they can carry, and only a copy of the family genetic guidebook to help them achieve success. Many will land close by, but spores of some plant diseases cross oceans looking for a new host. Others hunker down in soil or husks of dead plants and wait for spring.

Some basidios also have B and b mating genes. It's an unusual strategy to have two sets of mating genes, and to work, each pair has to match to allow mating. There are four main genders (AB, ab, aB, Ab), but different members of one species can also have different variants of their A/a or B/b genes. If you count each combination of variants as a different gender and do the math for some species, you end up with about 23,000 genders.[15] Hence, the flamboyant stories on the internet about fungi being exceptionally sexy.

The frequency of clones and the tendency to pump out thousands of asexual spores makes it quite likely that when you are on the hunt for love, you might encounter an identical twin. Imagine if every time you went out, you just kept meeting yourself. Your reaction would probably be "Oh no, not me again." Mating with your own clone is like mating with yourself; it's pointless because almost all genes are the same. A few fungi do go it alone because they are *A* and *a* at the same time; they clap their hyphae together and the sexual process begins. A few transgender yeasts switch back and forth between *A* and *a* during their life, but they still need a compatible partner if they are after sex.

The subconscious signal that leads you to your sexual partner is a chemical called a pheromone. Air, water, and soil are lush with small signaling molecules that elicit dramatic biological responses at very low concentrations. Your biochemical reactions, pathways, and by-products—the process known as *metabolism*—assemble and manipulate small molecules called *metabolites*. A common set of biochemical pathways hums along in the cells of all organisms. The process requires oxygen and chops sugars back into carbon dioxide, storing the released energy in a compound called adenosine triphosphate (ATP) that fuels the biochemical reactions of all cells. This is primary metabolism, or respiration. It is controlled by a core set of genes present in all organisms and powered within cells by tiny engines known as mitochondria. Metabolites produced in side reactions to the core biochemical cycles are secondary metabolites.[16] They are the chemical signals that you and your hyphae react to—pigments, flavors, smells, invitations, warnings. If you want to learn a second language, this is the one to study because it has the most native speakers: all life-forms, even predominantly visual animals like humans, use many of the same metabolites that fungi do. But individual species combine these chemicals in unique ways or interpret the signals differently.

As a socially aware fungus, you will take part in these noisy chemical conversations, broadcasting molecular signals to and receiving them from other fungi and microbes in the vicinity. These signals will often involve volatile metabolites that float through the air like gases or dissolve in water, often odorless or tasteless but not always. You might register common fruity or fragrant molecules called aldehydes or alcohols that often indicate the availability of sweet food. Or foul-smelling compounds that contain chlorine or sulfur, usually meant as a warning. Hormones tend to be larger targeted metabolites that transmit precise information within a colony or body or between individuals of a species. We call the hormones that affect behavior, such as an interest in mating, pheromones.

There is a purpose to all this communication, of course. You are not alone in the world. There are many of your own kind, fungal colleagues, as well as species of all the other kingdoms of life. You will be bumping hyphae with the mycelium of other species all the time. Other fungi and bacteria may want to share the same space or pass through openings in your webby colonies. You will have to get used to an intimate, colonial lifestyle. Some neighbors will help you out as long as you are not after the same food. Some will be downright nasty and aggressive. You need to figure out who is cooperative, who is competitive, and how to work with them in either case.

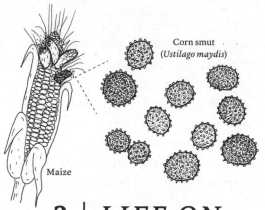

Corn smut
(*Ustilago maydis*)

Maize

2 | *LIFE ON*
THE COMMONS

From Mutualism to Parasitism
to Biological Invasion

THOSE OF US WHO LOVE THE OUTDOORS, whether we live in a city or the wilderness, know the pleasure of immersing ourselves in nature—the soaring trees, the industrious ants, the frogs bobbing about in the water, each with its own curious story. Nothing stands alone. When we retreat from urban bustle to wander in nature, we feel the cooler air flowing around us, the gentle breezes brushing against our skin. Soft light filters through the forest canopy, and quivering foliage mutes the cacophony of the city. Time slows. Dry, scratchy lichens cling to the trunks and branches of trees; ants haul chunks of dirt or vegetation about; squirrels hang mushrooms in the tree branches. When different species interact, an automatic interpretation that one is a friend and another an enemy doesn't cover the variety of relationships. Most of us seek out and rely on associates to make our lives work. Some are cooperative best buddies, but a few surprise us by thriving on competition and exploitation, while others sit in the middle, teetering back and forth from a little bit good to a little bit bad.

The term *symbiosis* (*pl.* symbioses) translates as "living together" and describes dependent or codependent relationships among species. Each participant is known as a symbiont, and when one grows inside another, the partner providing facilities is called a host. To meet the definition, at least one symbiont must depend on the relationship to complete its life cycle. This idea of symbiosis was popularized by Anton de Bary (1831–1888; if it helps you to remember him, think of him as Tony the Berry), usually considered the father of modern mycology. After a century on the fringe, the concept is now central to modern biology.[1]

Symbiosis is usually interpreted as cooperative: two allies are always together and mutually supportive, sharing nutrients or shelter. Some symbioses are extremely tight, with one or both participants completely reliant on their companion to survive. With time, signals of this codependency appear in the genomes or biochemistry of the partners. For example, one may shut down a biochemical pathway because its symbiont has a superior version. Other symbioses are casual and temporary. One or both of the associates might spend part of their life alone or keep company with other species. We are aware of symbioses that have participants from most kingdoms of life, but fungi seem particularly adept.

We often imagine symbioses as rare examples of biological altruism and focus on the doubly beneficial relationships called *mutualism*. But in the broad palette of symbiosis, a continuum exists between the cooperation of mutualists and the aggressive or destructive behavior of *parasitism*, where one symbiont harms or even kills the other. *Commensal* symbioses favor only one of the associates but don't harm the other. Many mutualistic or commensal partnerships slip over to parasitism when conditions change.

Traditional research on evolution emphasizes competitive relationships. Charles Darwin (1809–1882), for example, underplayed the significance of cooperation in nature. He portrayed natural selection as a battle between individuals over resources, with one emerging as the victor. He did not hold with win-win outcomes. Symbiosis is thus sometimes called "Darwin's blind spot."[2] When you observe behavior in nature, it's hard to draw a firm line between cooperation and competition. Any agreement between two creatures sharing resources is always under renegotiation.

Our surroundings are crammed with millions of microbes. When we consider this abundance, and all the interactions that might happen between any pair of species, the complexity of nature becomes clearer. Symbiotic relationships between fungi and other organisms challenge our preconceptions about individuality and the patterns of evolution.

Lichens: Mutualistic Fungus-Plant Symbiosis

Sudbury became famous for its lichens when I was a boy. The nickel smelters were the world's largest point source of sulfur dioxide, that gas notorious for its rotten-egg smell. The treeless terrain around the city was sprinkled with hundreds of unnaturally crystal-clear lakes acidified by pollution. After the Superstack became operational, remediation of the landscape began. Work crews strewed lime to reduce the acidity of the soils and planted tolerant grasses and shrubs. Teams of summer student researchers monitored buildings, rocks, and trees for the return of lichen species sensitive to sulfur dioxide. The changes in their abundance were early indicators that the contaminated landscape was recovering. Lichens are still used for monitoring air pollution around cities.[3] But what are these strange organisms?

Once you think to look around for them, lichens seem to be everywhere. If you explore along shorelines, you will see the spreading yellow and green crusts clutching on to rocks. Follow a path through the forest, and other lichens are glued to tree bark or festooned over branches. In cities, they spread on the stone walls and roofing tiles of old buildings. When you pick up a lichen, it is dry, tough, and leathery. The lichen body, known as a *thallus* (*pl.* thalli), is a dense felt of fungal hyphae. With a magnifying glass, the surface looks like a macabre landscape from a painting by Hieronymus Bosch. Monolithic columns and wobbly gall-like growths are littered over the crust. These are clusters or bundles of fungal hyphae coiled around algal cells. They break off like skin tags. Older lichens on trees or rocks often have a dribble of smaller thalli below them where these flakes dripped, then started to grow. Each thallus is layered: green strips of algal cells sit and harvest sunlight just below the fungal surface.

Lichens are the textbook example of mutualism.[4] They are usually portrayed as two-part organisms. One part is a mycelial fungus, the *mycobiont*. The second is a single-celled alga known as the *photobiont*, which can be either eukaryotic green algae or photosynthetic cyanobacteria (also called blue-green algae). About 20 percent of fungi, mostly ascomycetes, live only as part of a lichen and can't survive in nature without a photobiont. The mycobionts make flask- or cup-shaped sexual stages like other ascos, but when the asci shoot out ascospores, the spores need to find a new algal partner when they land. Only a few algae take part. Some of these can live alone but are seldom seen. And the algae don't have sex while entangled with the host hyphae, either because their partners repress them or because they prefer some privacy.

If we dissect lichens carefully, they are more complex than we expect. They are more like a community than an individual. Scattered pockets of yeasts or bacteria also seem to be symbionts within the thallus, but we don't know their role. Hyphae of other fungi often wind through older tissues, but it's unclear whether they are symbionts or squatters. The sum of a lichen is greater than its parts, and the collective survives in environments too harsh for the component species to tolerate on their own. In the give and take of their mutualism, lichens exemplify balanced trade. Mycobionts provide the shelter and the algae provide the photosynthesis. In lichens with cyanobacteria, the photobionts also capture nitrogen gas and transform it into ions used to assemble amino acids and proteins.

Next to mushrooms, lichens are the fungi that excite the most passion among citizen scientists because they are common, they are big enough to find easily, and they have an otherworldly appearance. Lichen-watching requires patience because many grow less than a quarter of an inch a year. Some favorite neighborhood thalli have been monitored by relays of observers for hundreds of years. Using a technique called lichenometry, they measure the growth rates of particular species and calculate the ages of individual growths. Lichenologists use digital photographs or trace patterns onto paper year after year. Even the retreat of glaciers can be estimated by the size of lichens growing on the rocks exposed on lateral moraines. One Arctic lichen was pegged at 4,500 years old.[5]

How can they survive for so long on rocks? Gel-like polysaccharides protect the hyphae and algal cells when the thallus dries, so it can survive long periods of drought. When

hydrated, salts and organic acids, like oxalic acid, ooze out
from root-like hyphal tendrils that anchor the lichen and,
along with physical forces of the weather, erode the outer
layer. When dry, the acids concentrate as microscopic crystals
along the hyphae. This is part of biological weathering, which
turns solid rock to powder.[6] The genomes of mycobionts give
more clues as to why some lichens live so long: they either lack
some genes that contribute to aging or switch them off.

Together, the fungus-alga consortium is particularly creative at producing secondary metabolites. Many of the about
eight hundred of these compounds studied so far are unique
to lichens. The fungal and algal partners have to be cultured
separately and are usually reluctant to reunite in lichen form
when paired in a Petri dish. By studying the biochemistry and
genetics of the separated partners, we can compare the out-
puts with those of an intact lichen to figure out the contribution of each to the symbiosis. The contribution of the algae to
the biosynthesis of the secondary metabolites isn't clear, but
cultures of mycobionts make different ones when they are
grown alone than they do in nature.

The pigments that give some lichens such a dramatic
appearance are secondary metabolites that act as sunscreens
to protect the thallus from ultraviolet radiation. They were
historically used as dyes.[7] The royal purple robes of modern
European monarchs are colored with the lichen dye orchil,
derived from species like *Roccella tinctoria*. The same species
is called the "litmus lichen." It provides the active ingredient
for pH indicator papers used in science classes that change
color when drops of acid or alkali touch them. Dyes for the
famous Harris Tweed brand of Scottish tweed cloth come
from *Parmelia* species commonly called "crottle" and are
extracted using human urine. (Tweed is often said to have a

distinctive smell, although commentators seem reluctant to describe it.) Vast leafy colonies of the lichens were gathered by the ton for centuries for their pigments from coastal rocks along the North Atlantic Ocean and Mediterranean Sea and are now in danger of extinction. Nowadays, the rare and expensive lichen dyes are mostly used by artisans and hobbyists interested in preserving traditional knowledge.

Folk healers layered lichens onto wounds, or prepared tinctures and administered them as drugs for hundreds of years before scientists discovered the antibiotic properties of moulds. Medieval and modern herbals offer instructions for using lichens or their essences to treat burns, coughs, consumption (tuberculosis), intestinal complaints, pneumonia, scarlet fever, or ulcers. Despite their frequent use in Indigenous and folk medicines, lichen products are rare in modern drugs. Chemical extracts of lichens contain many secondary metabolites with potent physiological effects, but not always pleasant ones.[8] The difference between a remedy and a poison—between a few molecules that fulfill one function and an overload that does something else entirely—is commonly learned the hard way. Usnic acid, the most abundant bioactive lichen compound, is added to personal care products like toothpaste, mouthwash, and deodorants. Its antiviral and anti-inflammatory properties in lab tests suggest possible drug development, although commercial preparations sold as weight-loss supplements were pulled from shelves after they were associated with liver failure. But some recently discovered lichen metabolites are being developed as possible anti-cancer drugs.

Beyond lichens, fungi and plants rub elbows in many mutualistic symbioses in nature, including in forests and on farms. Fungi are usually stereotyped as associates of plants, but thousands of species, perhaps the majority, are associated with animals, especially insects, some of them as mutualists. Because the structural carbohydrates of fungi and insects are both chitin, there is a biochemical logic to those liaisons and conflicts; the enzymes they have for making or breaking down their own walls or shells also affect the structural polysaccharides of their counterparts. These relationships between fungi and insects expose a fine balance between cooperation and competition, and also suggest that our definition of symbiosis has fuzzy edges. Must symbionts always combine in one body? Or can symbionts live side by side? If so, how does that differ from domestication?

Leaf-Cutting Ants and Their Mycelial Gardens: The Invention of Agriculture

On the northern tip of South America, nocturnal leaf-cutting ants sashay in conga lines along twigs, balancing ragged leaf fragments above them like windsurfers' sails. The leaves aren't food for the ants. They are compost—to nurture the gardens of delicious mycelium grown in the ants' underground cities. Sixty million years ago, ants started transplanting mushroom mycelium into their tunnels and caves. After thirty million years the ants and fungi came to trust each other, and the fungi settled permanently into the ants' gardens. With their invention of fungal agriculture, leaf-cutting ants expanded their territories and became some of the most elaborate insect societies, long before *Homo sapiens* evolved.[9]

After continental drift joined the American continents three million years ago, leaf-cutting ants crossed into what is now Mexico and the southern United States. When human agriculture started in the area about five to ten thousand years ago, the plant-plundering consequences of this ant-fungus partnership became well known. One day there was a lush garden and cool shade in the hut beneath the fruit trees. The next morning, the crop was gone, the trees were stripped bare, and the farmers were sweating and hungry.

The symbiosis works like this. Worker ants harvest the greenery and dance it over to the nests, where smaller ants called minims chop the foliage into flakes with their mandibles and shred them into paste. The minims then regurgitate onto their gardens and stack leaf fragments into a delectable mille-feuille, adding their own droppings as additional fertilizer. In response, the fungus offers up juicy, nutritious lollipop cells, which coat the nesting chambers of the hidden city. The queen embeds herself in a comfortable mycelial blanket in a secluded corner of the garden. After an adolescence of rampant promiscuity collecting and preserving sperm, she settles in for a couple of decades of egg-laying. The fat- and carbohydrate-laden fungal cells are manna for the larvae emerging from her eggs.

Nowadays, about two hundred species of leaf-cutting ants select the mycelium of one of four tasty basidio species for their gardens. In the most intricate of these symbioses, the chosen fungus is *Leucocoprinus gongylophorus*.[10] For a colony to stay at its peak fitness, the ants try to maintain a pure clone of their fungus. Brawny soldier ants provide security along trails and inside the nests to repel unwanted visitors. Waste is gathered and piled up in different chambers. Worker ants are barred from moving from one garden room to another to reduce the chance of cross-contamination. And if a weed fungus sneaks in, the minim patrols detect it and dispose of it. If one of the gardens in the multigalleried colony fails, it is sealed off. This vigilance ensures the health of the ants: fungi are all they eat.

New nests and gardens are constantly excavated and constructed, attached to the established colony. Worker ants drag soil up, one grain at a time, and dump it on the surface. Brazilian biologists curious about the extent of the underground cities buried beneath a meadow of tiny, sandy volcanoes poured 10 tons of concrete into the openings on the surface. Three weeks later, backhoes and student volunteers removed the surrounding dirt, revealing a metropolis resembling a demented carnival ride designed by Dr. Seuss. The structure was 25 feet deep, about 500 square feet in area. Hundreds of soccer ball–sized fungus gardens and waste facilities were connected by a chaotic network of curved tunnels used for ventilation and transportation.[11]

A queen might reign over this domain for twenty years and count up to eight million ants as her subjects. When a new queen hatches, matures, and sets off on her nuptial flight, a coterie of males accompanies her, cradling hyphae of her mother's fungal clone to seed the new colony. A queen's death marks the end of her metropolis. Upon her demise, as if in royal tribute, a mushroom stalk rises through the remains of the abandoned nest, spreads its cap, and releases basidiospores that swirl off into the air in search of a new monarch. Apart from this instinctive response to escape the empty nest, these fungi seldom bother making mushrooms anymore.

Other microbes are part of this community. Some bacteria in the garden capture atmospheric nitrogen and transform it to meet the dietary needs of the fungus and the ants. A weedy mould called *Escovopsis* overgrows some gardens, turning them into a collapsed, sandy brown mess. It is a parasite of the mutualistic symbiosis and relies on it; it is never seen anywhere else. In some nests, worker and queen ants sport colonies of mould-like bacteria, called actinomycetes, on their backs. These *Streptomyces* species look like white splotches of paint and produce antifungal antibiotics that repel invasive moulds, including *Escovopsis.*

Domestication can be interpreted as one of the many flavors of symbiosis. The word's definition—"belonging to the house"—gives naming rights to the partner that provides the infrastructure, but the process is usually a joint endeavor. Strains that are isolated long enough from their original wild populations undergo a kind of genetic Stockholm syndrome. Their genomes alter in ways that ingratiate themselves to their hosts. If the host also changes in response, the result is symbiosis. Ants did such a good job of predigesting leaves that the fungus got rid of its cellulase enzymes. The ants, in turn, stopped making their own supply of the amino acid arginine because it is so plentiful in the fungal cells.[12] Leaf-cutting ants initiated the interaction with fungi that they cultivated, but the relationship evolved from domestication to codependency. Domestication is a recurring theme with fungi. It leads to some interesting twists when we reflect on our relationships with agriculture and food.

These two examples of mutualism—lichens and leaf-cutting ants, both systems enhanced by third-party microbes—complicate how we look at ecology and evolution. More questions arise when we consider other symbioses that sit farther from the balance of mutualism.

Commensalism: Friends Without Benefits

Commensalism is a type of symbiosis halfway along the spectrum from mutualism to parasitism. True commensalism is rare, and it is also often short-lived. Some fungus-insect interactions inhabit this middle ground. Trichomycetes, or trichos, are one example, a group of zygomycetes that live in the intestines of millipedes, crustaceans, and the wriggling larvae of aquatic insects like caddis flies, where they absorb leftover food from their hosts. DNA studies revealed that some trichos are related to amoebae and not fungi, so "trichomycete" is now an informal nickname instead of a formal classification.[13]

Many insects pass through several phases during their short lives, and their digestive tract is renewed with each metamorphosis. Trichos, for example, usually only colonize larvae. Inside their host larvae, the trichos' finger-like hyphal anchors grasp the walls of the hindgut to stop the mycelium from being flushed out. Then the trichos' sausage-like hyphae float in the gastric juices and sop up secondhand nutrients that dribble towards the exit and would otherwise be waste. This commensal relationship ends when the larva transforms to its next stage.

Commensalism illustrates just how fluid symbiotic relationships can be. Some trichos might leak a few of their own sterols or B vitamins to enrich their partners, and then they are modest mutualists. Others, like the mosquito fungus *Smittium morbosum*, accidentally kill their hosts, not through aggression or overt pathogenicity but by stopping the larvae from molting. The fungus penetrates the inside of the mosquito larva's gut and attaches to a chitinous part near the exit. The larva keeps growing inside its exoskeleton, but when it outgrows the available space and starts to molt, the fungal attachment interferes with the separation of the old and new exoskeletons and the expanding larva crushes itself to death. Some scientists wonder if this *Smittium* might be applied to kill larvae of the mosquitoes that carry malaria.

Smittium is an example of a commensal relationship that becomes parasitic as its relationship with its host continues. Many symbionts lurk until they can shift the situation to their advantage.

The Gray Margins: Parasitic Symbionts or Pathogens?

The link between disease and microbes is usually called germ theory. The credit for this idea often goes to Louis Pasteur (1822–1895), who demonstrated that bacteria cause postpartum infections in humans. Diseases caused by symbionts are considered parasitism and are at the opposite end of the symbiotic spectrum from mutualism. A parasite grows inside its host and absorbs nutrients or energy without offering any benefit in return. Some parasites start off as mutualists or commensals, and then when the host ages or the environment changes they alter their strategy and start to weaken their host. The distinction between a parasite that weakens its host from within and a pathogen that attacks from outside might seem semantic. But if you are the one suffering disease and asking, "Why did I get sick?" the difference is important.

After Pasteur became famous for his breakthroughs with human pathogenic bacteria, he studied a parasitic disease of silkworms called pébrine, or pepper disease. It is caused by a single-celled organism called *Nosema bombycis*, one of fourteen hundred species in the poorly understood phylum Microsporidia, which DNA studies confirmed as part of the fungal kingdom only over the last decade. Pébrine attacks silkworm larvae and spends its whole life in the cells of its host. Like most microsporidia, it unravels a coiled whip-like filament inside its host cell, jabs it through the membranes of neighboring cells like a needle, and then injects a nucleus and some cytoplasm through the channel to initiate a new parasitic cell. The insect is greatly weakened as the parasite takes over its tissues and saps its energy, and the disease is often fatal. Survivors are often too weak from constant vomiting to spin silk, or they weave only poorly formed cocoons. The dark spores speckle the larval tissues as if they have been sprinkled with pepper, and they are passed from one generation to the next in insects' eggs. Pasteur tried to convince reluctant farmers to sort out and discard infected eggs to develop disease-free stock; their skepticism was remedied by the charms of Pasteur's youngest daughter, who showed them how easy it was to use a microscope. Pasteur was then credited with saving the French silk industry.[14] *Nosema* returned to public attention more than a hundred years later when it caused infections in people with AIDS.

Several fungal parasites of plants, often labeled "obligate biotrophic pathogens," depend on living plants to survive and complete their life cycles. In hot or humid weather, hyphae of a group of ascos known as powdery mildews spread over the top and bottom surfaces of green leaves. Short microscopic needles penetrate the host epidermis and siphon nutrients into the hyphae, weakening the plant so much that its leaves wither and mature seeds may not form. Over the summer and early autumn, the tissues get covered with the white, powdery chains of barrel-shaped asexual spores that give the mildews their name.

In vineyards, protecting delicate vines from grape powdery mildew (*Uncinula necator*) has been a long-standing priority ever since it was accidentally introduced to Europe in the early 1800s. In the nineteenth century, the parasite reduced grape harvests in France by up to 75 percent. Often the grapes were overgrown and did not fill out. The diseases caused by some of these species were the inspiration for the first antifungal pesticides. To discourage passersby from snacking on grapes, winemakers of the time often sprayed vines along roadways with a bitter, lurid green slurry of copper sulfate and slaked lime. The French mycologist Pierre-Marie-Alexis Millardet (1838–1902) noticed that these grapes rarely suffered from mildew. He experimented with fogs and sprays and developed the first chemical fungicide, known as Bordeaux mixture. It suppressed mildew effectively but turned the skin of fruit pickers blue. This mixture is still tolerated by some organic growers because its components are considered natural rather than artificial. Grape plants, of course, were transplanted all over the world as the European taste for wine spread. And powdery mildew sometimes traveled with them.[15]

Quite a few fungal diseases are difficult to pigeonhole as parasitic symbionts or infective pathogens. Smuts are microscopic basidiomycetes that attack plant ovaries, especially those of grasses, and replace the tissues with masses of hyphae and dark brown spores. The common name "smut" indicates their resemblance to soot and in no way suggests the more sordid definition of the word. Corn smut attacks about a third of maize grown in the world and is especially brutal to sweet corn. But to plant pathologists, corn smut is a fungus with saprobic, pathogenic, and symbiotic phases in the same life cycle. For part of its life, *Ustilago maydis* is a saprobic yeast growing in soil and on crop debris. The yeasts splash onto the plant and infect it as a pathogen, but then it transforms into a hyphal form. The hyphal form completes the sexual cycle, and because this stage depends on the living host, it is considered a parasite. It meanders through the ear silk into developing kernels and then grows until it bursts out as galls that look like coal dust. The galls make so many powdery spores that combines trundling through heavily infected crops are followed by billowing clouds that make their engines appear on fire. Most farmers consider corn smut a scourge; however, Mexicans have had a taste for immature smutted corn, which they call *huitlacoche*, since the time of the Aztecs. Rich in the essential amino acid lysine, the velvety paste adds an earthy, smoky note to tortilla dishes and soups and is said to reduce cholesterol. The price of fresh or tinned diseased kernels far exceeds the profit from healthy cobs.[16]

When we think about microbial diseases, though, most of us think of pathogens, which attack a host from the outside, weaken or kill their victim, and then look for a new host. Many pathogens spread quickly and broadly and there is nothing symbiotic about them. They are out for themselves, and we need to beware.

Invasion: The Aliens Have Landed

Look in your yard. Many of the plants, birds, rodents, and other animals outside your windows came from somewhere else. Some, like house sparrows and dandelions, were carried overseas deliberately, then freed or planted by settlers to remind them of home. Others—the house mouse, earthworms—came along unnoticed. Whether they were welcome passengers or stowaways, organisms transplanted from their native lands to a new environment are called introduced or non-native species. Many domesticated crop plants and livestock animals have been with us so long we consider them natives, but most are non-native species. They often carried their diseases, pests, and symbionts along with them.

Biological invasions happen when non-native species explode in numbers and push out or kill local, endemic species. This destabilizes the ecology of established communities. If the disrupted organisms are economically important or affect human health, the consequences are severe. Ships sailed back and forth for hundreds of years between North America and Europe before anyone wondered whether shuffling plants and animals around might be a bad idea. But in hindsight, ever since the black plague epidemics of the fourteenth century, which were caused by the bacterium *Yersinia pestis* riding along on fleas hidden on rats that were hitchhiking on ships, the importance of preventing biological invasions has been clear.[17] Fungi are often a guilty party, or accomplices in the crime.

The most famous invasive crop epidemic is late blight of potato, which Irish tenant farmers called murrain, caused by the fungus-like organism *Phytophthora infestans*. Although it looks and acts like a fungus, the evolutionary history traced by its DNA classifies *P. infestans* as a non-photosynthetic alga. The Irish love their potatoes, which were introduced to the Emerald Isle from Peru in about 1600. In the mid-nineteenth century, Irish farmers grew wheat for their British landlords and planted potato crops on the side to provide for their families. The tubers stored well through the winter, and an acre of stony soil could feed a family of six.

However, the summer of 1845 was mild and wet, and the murrain, which had first been seen in France the year before, flourished. Politicians were certain that newfangled electrical wires passing over potato fields were the problem. The English clergyman Miles Joseph Berkeley (1803–1889) advanced the outrageous claim that a microbe was the cause. With his microscope, he saw candelabra-like hyphal structures dangling from the underside of the infected leaves. Spores from these structures drifted off on the wind in search of healthy leaves. But if they landed on damp dirt, instead of hyphae growing out, they cracked open and a swarm of tadpole-like zoospores swam off through the soil and infected fresh tubers. Healthy plants collapsed into putrid, stinking masses within days of infection as hyphae of the *Phytophthora* sopped up the starch. Stored tubers rotted into black, mushy lumps, their nutritional value destroyed.[18]

A million Irish people died from starvation and disease between 1846 and 1851, with 1847 by far the worst year. Two million more fled the country.

The murrain was always considered an invasive disease in Europe. In 2015, genomic analyses confirmed that the virulent strain originated in the United States and crossed the Atlantic in infected tubers. It remains an invasive pathogen, still causing losses of $5 billion per year. It is now a bigger problem on tomatoes. Starting in 2009, severe blight destroyed seedlings sold to home and organic gardeners across North America. They were infected by a virulent clone from Europe that jumped back across the Atlantic.

Often, the aggression of invasive species follows their escapes from their natural enemies and competitors (and sometimes symbionts) in their original terrain. To combat the invasion, biologists sometimes import these natural adversaries from the home territory with the hope that they might slow the invasion down. A deliberate and (we hope) controlled, beneficial biological invasion that is used to combat a harmful one is known as biological control, or biocontrol. And the biocontrol agents—biopesticides, or biological pesticides—are usually considered environmentally benign alternatives to chemical pesticides to reduce or eliminate damage to plants or animals caused by insects, fungi, or other pests.

Biocontrol had a rocky beginning. In the 1930s, some 62,000 American cane toads were introduced into Queensland, Australia, to reduce exotic beetle infestations in sugarcane plantations. Unfortunately, the feral toads spread across much of north and east Australia and, over the next fifty years, caused serious damage to native ecosystems. So the worry with any biocontrol agent is that, like the cane toad, it might settle in permanently, spread, interfere with endemic species, and become an invasive species itself. However, *Puccinia* species and other rust fungi tend to attack only specific hosts, and deliberately introducing such a fungus to chase down accidental invasions by weeds is sometimes effective. Populations of the targeted plant are usually significantly reduced, although seldom are they completely eliminated. But if the unwanted weed is severely weakened, the epidemic caused by the pathogen also fades away, as long as it doesn't jump to another host.[19]

The Invaded Cell

Biological invasions happen frequently, but most are minor and don't endanger entire ecosystems. And the consequences are not always bad. Sometimes they are so beneficial that they lead to dramatically new ways of life. The best-known example involves a bacterium, with all eukaryotes as the beneficiaries.

About 1.5 billion years ago, mutualist symbioses among single-celled organisms set the stage for the rise of eukaryotic life. Every cell inside every plant, animal, fungus, or protist has nuclei with the organism's own chromosomes and DNA. But all eukaryotic cells are also chimeras.[20] Mitochondria are essential intracellular organelles that produce all of our energy. They were originally free-living bacteria and were engulfed by or invaded our ancient ancestors' cells, then settled in permanently as symbionts. Without them, none of us would be alive. Mitochondria have their own DNA and their own small genome. During sexual reproduction, when the chromosomes in our nuclei merge and rearrange, and during routine cell duplication when our nuclei divide, the genomes of our mitochondria multiply on their own. In animals, the mitochondrial genome that passes through the generations comes from the egg cell of the mother. Thus, our mitochondrial symbionts allow us to trace our maternal genealogy, but not our paternal one.

There's more. About a billion years ago, plant cells absorbed photosynthetic bacteria that also moved in for keeps; we call these symbionts chloroplasts. The capture of energy from sunlight by plants and the management of energy in our cells both rely on symbiosis. If our symbionts decided to leave, we'd be in big trouble.

Symbiosis forces us to reexamine many of our assumptions about how nature works. The idea that organisms live inside one another, all over each other, and beside each other, as friends sometimes and enemies at other times, can be difficult for humans to fathom. We like the simple and somewhat romantic idea of mutualism: two unrelated organisms moving in together, sharing a life, the bliss of two melded into one. But many symbioses are also transient or harmful. At the same time, fungi are great collaborators, and these relationships are a vital part of the success of many ecosystems. These behaviors and codependent relationships can change over time and with age: many symbioses have a beginning, middle, and end. It is in forests where fungi's diverse talents as symbionts, decomposers, nutrient cyclers, and chemical communicators are on full display. Exploring there, we will have our first clear understanding that nature as we imagine it is different from nature as it actually is.

PART 2

THE FUNGAL PLANET

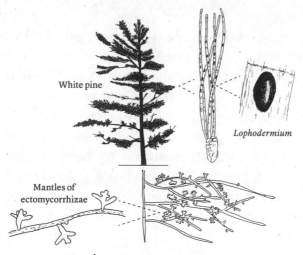

White pine

Lophodermium

Mantles of
ectomycorrhizae

3 | *FORESTS*

Seeing the Fungi for the Trees

FEW OF US have experienced an unlogged forest. When you step into an old-growth grove, one of the few that was never logged, the mature trees might be hundreds of feet tall. Embracing a trunk is a team activity. The overlapping canopies form a vaulted ceiling, casting shadows onto the understory. But if you look closely, you will see that the branches of those neighboring trees seldom touch, a phenomenon called crown shyness. We don't know why this happens, but it may prevent some pest insects or parasitic fungi from jumping from one tree to the next.

What we think of as wild forests are often scrappy affairs where older trees were removed long ago and new generations have regenerated on their own in a haphazard way or been replaced by rows of seedlings. Commercial forest plantations are intended to grow quickly so that lumber can be harvested and new trees planted. In many parts of the world, the cultivated seedlings are non-native species. North American trees like Douglas-fir and radiata (or Monterey) pine have been transplanted around the globe in temperate climates, and subtropical areas are often filled with plantations of cloned, fast-growing gum trees (*Eucalyptus*). Many ornamental trees planted along boulevards or in urban parks were also carried across oceans along with their symbionts and pathogens. These human-made woodlands are designed ecosystems—fake ecology, if you will—planned by evaluating which trees and shrubs might work together in particular climates and soils. They may still be productive and lovely, of course, but the complexity and resilience of a natural old-growth forest is difficult to duplicate.

Fungi are critical partners in the eventual success or failure of all forests. No matter how anemic the woodlot, in the autumn you will find mushrooms throwing their spores into the wind like confetti at a wedding. But everywhere you look in a forest, no matter what the time of year, you are looking at fungi—they are hidden in plain sight. They cover the outside of the leaves and bark, hide inside leaf tissue, or wind their hyphae around roots. They decompose dead pockets of wood. In the soil, saprobes scavenge minerals and break down plant litter. On the surface of the ground, fungal colonies are busy spreading, sending mycelial fans or cords through the leaf litter, sometimes dispatching shoestring-like rhizomorphs in their quest for more territory.

Forests are like cities. The trees are like buildings connected by hidden infrastructure, elaborate transportation routes, and large populations of all kinds of organisms going about their own business. Many fungi are busy outside, but many buildings are also packed with tenants.

The City in the Sky: Fungal Endophytes and Epiphytes

Outside my window in eastern Canada, white pines jut up through the canopy of maples and cedars like crooked towers on a skyline. A tree is like a high-rise apartment building. Its foundation may run deep but a lot of excitement happens above ground, in rooftop restaurants and bars, with hundreds of stops at floors in between. Birds and insects cruise around, and crowds of tiny mites, millipedes, worms, and mice party in cavity apartments. Balcony gardens of aerial plants, mosses, and lichens droop over the boughs. Some inconspicuous moulds are among the *epiphytes* that grow on the surfaces of plants. They creep over the waxy outer layer of leaves and bark, plop out the odd spore, and sponge up nutrients that dribble out of the plant.

Despite their evergreen reputation, pines do drop their needles—but a few at a time instead of all at once, usually after two or three years of hanging on. Sometimes a bundle of needles remains aloft, caught up in a branch. The rest sprinkle down into a slippery carpet that accumulates on the forest floor. Each fallen needle starts off smooth and golden, but after a few weeks black splotches the size of a pinhead appear. When magnified, the splotches look like pairs of dark lips and when they get a bit wet, they open to reveal a firm, whitish gel. Black rings appear around some needles, as if someone has used a pencil to separate the different sets of lips. We used to assume that this fungus, which is called *Lophodermium* (or lopho for short), grew into dead needles after they hit the ground, or that it was a pathogen that caused leaf cast or needle blight disease. But over the past few decades, these impressions have changed.

We've known about *endophytes*—organisms growing inside plants—for more than 150 years.[1] The term was coined by Anton de Bary, the same man who brought symbiosis to light in the mid-1800s. Until recently, because the fungal species involved could not be identified and grew so slowly, endophytes were a black box. We imagined they were symbionts of some sort because they didn't seem to cause disease and host trees seemed to tolerate them. But what are these fungi doing? With a microscope, you can see sparse hyphae between the cells of the plant in the needles, leaf buds, and bark. The plants seem to regard this as routine. It seems clear that the transaction must involve fungi soaking up nutrients flowing through the living tissues, but what do plants get in return? Endophytes seem to give trees some flexibility in coping with environmental changes. Trees with vibrant endophyte populations grow taller and faster, tolerate drought, and seem less sensitive to toxins like heavy metals.

At the first scientific conference I attended as a graduate student, the keynote speaker was an eccentric retired professor who showed photos of slow-growing fungal cultures he teased out of living conifer needles. They were black, coal-like blobs that crept across the agar in his Petri dishes at about one-sixteenth of an inch per month. The blobs stopped after a while and sat there—it was hard to tell if they were alive or dead. The prof muttered into his lichen-like beard that this slow growth was why he had so few grad students. With only five years to do their research, they would never be able to grow enough mycelium to use for experiments. And because most of his cultures made no spores, no one could identify them—and how could anyone write a thesis about a fungus they could not name?

Isolating these cultures was hard work. Conifer needles are messy on the outside, covered with spores of epiphytes, or of saprobic moulds, that float around in the air or scuff off insect feet. To isolate what is really growing inside the needles, you have to clean the outside, a process called surface sterilization. This means a lot of washing with sterile water and some soaking in alcohol or bleach. You want to kill off loose spores of any outsiders but keep the insiders alive. It takes some tweaking to get this process right.

Once DNA barcoding came along, we were shocked to discover that among all the cultures and DNA barcodes directly retrieved from plant tissue were thousands of species of endophytic fungi. They grow in almost all kinds of plants. Some are related to species that we'd seen before elsewhere in the forest as free-living saprobes or parasites, but others were always concealed from our sight and lacked known close relatives. Most, including lopho, are ascos.

Now that we can detect lopho by its DNA and don't have to wait for cultures to grow, we know that it occurs in almost every one of the several million living green needles on a healthy pine. Lopho spends most of its life hidden inside the foliage.[2] The tough waterproof wall provided by the epidermis of pine needles protects the hyphae growing inside from the sun and drying winds. The pencil-like lines circling the colonized needles are called barrage zones. They form in many kinds of leaves and needles and are a sign of multiple clones or species competing for the same tissue. Sometimes, within such spaces, squabbles break out with the endophyte next door. Feuding neighbors try to hoard nutrients for themselves and build walls of dark cells at the edge of their territory to block out other hyphae from competing. The patterns created in lumber by barrage zones attract artists and furniture makers, who call the phenomenon spalted wood.

The canopy of a tree is full of families, friends, strangers, and enemies all relying on one landlord to provide sustenance.

When things go well in this symbiotic high-rise, all the tenants pay their bills and contribute to maintenance and everyone prospers. But there will always be a few who make a mess, start an illicit operation in a back room, or get too aggressive towards the neighbors or their host. In other words, some endophytes slink away from mutualism and become parasites. They are already comfortably positioned inside the host's cells, and the tree's defense systems are focused elsewhere. The unexpected betrayal can be fatal for the host.

An inconvenient truth is that many endophytes, including some lopho species, cross the line from beneficial to harmful often enough that some trees would rather not have them around. A mature tree might have hundreds of different endophyte colonies. As long as most are young, vibrant, and on average have a positive influence, the host keeps them around. But eventually the landlord needs to issue eviction notices to wayward tenants before the whole building collapses. This is one reason that trees, including conifers, drop their leaves.

For lopho, it is only when the needles fall to their demise that the fungus marshals its resources to make asci and ascospores. The lip-like dots are its sporulating bodies and the white gel inside is the layer of water-cannon asci that fire ascospores into the air. Their targets are the tender young needles of germinating seedlings poking out of the undergrowth. Asci produced on needles hung up above ground have a clear shot at young buds in canopies of nearby mature trees. Every spring, at the same time that new needles sprout during the active period of branch growth, the trees are pelted with lopho's ascospores trying to ingratiate their way back inside. The mixture of symbionts in any one plant is thus constantly changing. This is not the same kind of stable, long-term symbiosis between two partners that we see in a lichen. With so many endophytic fungi, and so many hosts, there is a constant reshuffling of micro-symbioses, each vacillating between mutual benefit and antagonism, exploitation, or indifference.

Other factors might be at play, but the improved health of plants associated with endophytes may follow from reduced insect damage. Trees often suffer from moths or beetles eating or drilling their way in, laying their eggs, and rearing their broods using living bark or leaf tissue as food. They may not kill the plant, but they weaken it, reduce photosynthesis, slow growth, and affect the efficiency of systems that take up and distribute water. In its secure shelter inside a needle, an endophyte has the same interests as the host—it doesn't want to be chewed up by bugs or elbowed out by other fungi. If it can produce toxins that inhibit pests but don't affect the tree's biochemistry, this helps them both. Insects tend to avoid tissues colonized by endophytes. Some cultures of these fungi pump out toxins that either kill insects or make the leaves less palatable to voracious larvae. Other compounds, or sometimes the same ones, also repress pathogenic fungi.

With so many fungal endophytes in trees, it's unsurprising that different species have different schemes for colonizing or recolonizing plants. When trees adjust to the change of seasons, their leaves or needles start shutting down and get ready to fall off. Some endophytes send out spores while they are still in the canopy. But many endophytes spend only a part of their life span inside plants. They ride the leaves as they drop to the forest floor and switch to a saprobic mode. They consume bits of dead wood or discarded leaves or send off acrobatic, star-shaped spores to float down streams. Some endophytes are very fussy about which tree or plant species they engage with in symbiosis, but less particular about which they will consume on the forest floor. Eventually, branches, needles, leaves, flowers, seeds, and nuts all tumble to the ground, carrying their fungi with them and exposing themselves to those waiting in the soil. Below, in the subterranean city, the forest is also a busy place.

Notes From Underground: Mycorrhizae and Rhizosphere Fungi

In human cities, buried foundations keep buildings from tipping over, and networks of electrical cables and plumbing pipes connect them to their communities. Forests are similar. Roots hold trees upright and provide an intricate biological infrastructure to keep water and nutrients pumping from the ground to the canopy. For a long time, botanists assumed that trees did all this by themselves. Now we know they have a lot of help.

You'd need a magic spade to cut a clean, vertical slice through the soil and woody roots below a trunk. If you had one, you'd get a clear view of stratified layers of organic debris and soil and see how extensively fungi interact with trees. In the rich, loose soil near the surface in a forest, you often find whitish sheaths surrounding the spreading rootlets of nearby trees. The root system of any one tree has tens of millions of these spongy fungal knobs at the ends of its youngest, finest roots. They are called mantles. Mycelial clouds fan out from them, hyphae and hyphal bundles arcing outward into the soil like shafts of lightning.

Mantles were first noticed on pines in 1840 by the German professor Theodor Hartig (1805–1880). With his microscope, he observed a sparse network of canals between cells of tree roots, like threads navigating a maze. It was another thirty-five years before they were recognized as fungal hyphae. The lattice was named the Hartig net. By patiently following hyphae from a mantle to a mushroom and then back to a different mantle, you can trace hyphal connections between roots of different trees. The network is so fragile and the hyphae are so inconspicuous that this process requires special sensitivity—not green thumbs, but fungus thumbs.

The fungal nubs on pine roots seemed like a novelty until mantles and Hartig nets were found on many different European tree species. Most foresters assumed that the fungus must be parasitic on the roots. But German botanist Albert Bernhard Frank (1839–1900) called the associated fungal mycelium a "wet nurse" for the tree, suggesting that it controlled uptake of nutrients into the plants. He also gave it the name *mycorrhiza*, meaning "fungus root" (from the Greek words *mukēs* for "fungus" and *riza* for "root").[3] In nature, these fungal species are always associated with their symbiotic hosts; they don't have an independent, saprobic form like some endophytes do. Although trees can survive without mycorrhizae, they grow faster and taller with them, and the symbiosis seems mutually beneficial. However, some ecological accountants insist that fungi get more from the relationship than plants do and could be interpreted as mild parasites.

Mycorrhizae are the star symbioses of forests. They have evolved on about sixty different occasions since life first colonized land, and modern plants are heavily invested in having fungi colonize their roots. Only 8 percent of land-dwelling plant species lack some kind of mycorrhizal associate.[4] There are several different patterns. Orchids don't have chlorophyll when they are young. Their mycorrhizal fungi, related to the basidios called jelly fungi, provide nutrition until the orchid starts its own photosynthesis. They are so important that most orchids package the fungi into their seeds. A few plants, like the ghost pipe, *Monotropa uniflora*, no longer make chlorophyll at all and are parasites on their mycorrhizal fungi. Mycorrhizae of blueberries, cranberries, and their relatives, which tend to grow in poor soils, are usually partnerships with ascos. These so-called ericoid mycorrhizae are essential for the nitrogen nutrition of the plants. The most ancient fungal root symbionts are arbuscular mycorrhizae, symbionts of many agricultural plants.

The mycorrhizae that form mantles and Hartig nets in forests are called *ectomycorrhizal* fungi, because of their extensive hyphal development outside (= *ecto*) the roots. The hyphae involved in nutrient exchange grow between but do not penetrate the root cells. About eight thousand fungal species, mostly basidios, take part in ectomycorrhizae.[5]

Many of the mushrooms decorating the forest floor are the spore-producing bodies of ectomycorrhizal fungi: the smooth, shiny, rubber ball–like caps of *Russula* species, the colorful *Lactarius* milk cap species with their bleeding gills, the deadly destroying angels (*Amanita*), and most of the boletes, for example. Some of the best edible wild mushrooms are also ectomycorrhizal. My favorites are chanterelles, *Cantharellus* species such as *C. cibarius*, with trumpet-shaped orange mushrooms that smell faintly of apricot and have blunt, thickened ridges instead of sheet-like gills. Their flavor is nutty and fruity and they have a pleasantly firm, fibrous texture. You can buy dried ones but they're not quite the same. Despite the financial incentive, nobody has figured out how to grow them in controlled conditions; they need the connection to living trees, and perhaps other prerequisites that we don't yet understand.

Most fungal species tend to be symbiotic with just one species or genus of tree, giving mushroom hunters valuable clues about where to look to forage a meal. Other mycorrhizal species are more casual, perhaps favoring either coniferous or broad-leaved forests in general. Tree species are more promiscuous than fungi. Douglas-fir roots associate with as many as two thousand fungal species, though usually only ten or twenty different fungal species form ectomycorrhizae with a single tree.[6]

Mycorrhizal networks in forests capture the imaginations of nature lovers and scientists. Hyphae of different species can't fuse directly, but they can connect to the same roots as other mycorrhizal fungi. A vast hyperbranched root and hyphal framework links all the trees in a forest into a unified network. It is not a centralized neurological system like some animal brains, but a distributed, multinodal network like the cabling used to distribute electrical energy and information among buildings. This presumptive communication grid is often compared with the internet and has been dubbed the Wood Wide Web.[7]

What kind of communication passes through these root-hyphal networks? The first experiments were done in the 1950s by Elias Melin (1889–1979) and Harald Nilsson (1921–1989), two Swedes working at the famous university in Uppsala where Carl Linnaeus, the father of modern taxonomy, once taught. They grew pine seedlings in cone-shaped Pyrex flasks and coaxed them to form mycorrhizae with cultures of boletes that were sequestered in glass cups. They then added radioactively labeled compounds to the cups. When they tested the needles of the seedlings later, the scientists detected radioactive forms of carbon, phosphorus, and other minerals that could only have passed into the plants through the Hartig net and the hyphal connections between roots. In return, the researchers found, sugars made during photosynthesis in the leaves moved from the roots into the hyphae. Minerals and nutrients flowed in both directions.[8] It is as if the plant delegates some critical root functions to mycelium, accepting that the fungus is better at it than the plant itself.

Recent studies in natural forests show that radioactivity from carbon dioxide tracers released near one tree shows up in nearby trees, sometimes of different species, in less than an hour. The first tree absorbs the gas and transforms the labeled carbon into other molecules. These metabolites travel into the roots and the hyphal network, ending up in the tissues of a neighboring tree. The node (or mother) trees connected to the most other trees are often the tallest and oldest in a forest. They also share more nutrients with their own seedlings than they do with others, suggesting that they recognize and favor their offspring and can adjust the flow through the hyphal network. While we know signal molecules are exchanged through mycorrhizal networks, so far we don't know much of what they are talking about. It is probably mostly gossip about the weather or aggressive insects who have moved into the area.[9]

Mycorrhizae are a win-win transaction for trees and fungi, and because they improve the productivity and health of forests they are important to us too. Even in forests disrupted by logging or disease, mycorrhizal fungi can be of biological significance and economic value. For example, the mycorrhizal pine mushrooms *Tricholoma matsutake* and *T. magnivelare*, known to cognoscenti by their Japanese name matsutake, tend to grow in degenerated forests. On the left coast of North America, the Wild West is reborn each year as gaggles of mushroom pickers migrate northward in their pickup trucks, following the seasonal fluctuations that favor matsutake. The mushrooms are trucked out of the wilderness and loaded onto 747s to be sold in markets in Japan, Korea, and China within a few days.[10] They are used more as a flavoring agent than as a vegetable—a few thin slices add an aromatic, spicy note to a bowl of soup. It is the piney, meaty, floral bouquet that makes this mushroom so popular in Asia. Describing the fragrance is a challenge for even the most nuanced haiku master, but it is said to "recall times past."[11] Westerners often appreciate these subtleties less. For a long time the matsutake that grew in Europe was called *Tricholoma nauseosum*, and its odor was described in field guides as being like dirty socks.

The aromatic black French Périgord truffles (*Tuber melanosporum*) and the Italian white truffles (*Tuber magnatum*) are lifelong mycorrhizal symbionts of ancient oaks. They grow underground, where it takes decades for seedlings to develop mature mycorrhizae that produce ripe truffles (the sporulating bodies). They resemble distorted lobed potatoes or chunks of coal. These ascomycetes started out making platter-shaped disks, but over evolutionary time they folded over on themselves and started making asci inside. Because they could no longer blast their ascospores into the air, they needed to find another way to disperse. Their intoxicating smell assures their survival, attracting rodents, birds, and swarms of flies to carry off the spores. The underground tunnels of some voles lead from one truffle to another. Dogs and pigs can be trained to sniff them out. Early claims that truffles' musky odor came from the same sex hormone (5α-androstenol) produced by pigs and male humans didn't stand up. The smell is a complex mixture of volatile organic compounds; chemical

analyses identify more than one hundred metabolites that
contribute to the odors of various truffles. The odor is primarily of a small sulfur-containing metabolite called dimethyl
sulfide. The same molecule contributes to halitosis and gives
overcooked cabbage its distinctive "atmosphere." Yet truffles routinely sell for more than $900 a pound, and flawless
samples weighing two to three pounds have netted more than
$100,000 at auction. Buyer beware, however—the truffle
trade is notorious for fraud.[12]

Most mycorrhizal and endophytic symbioses tend to be
short-lived. Any one tree has many fungal partners in both its
roots and its foliage. Those relationships shift as a tree ages
and its environment changes. Despite this constant fluctua-
tion, trees, like ancient buildings, maintain their stability.
They are chimeras of their own plant bodies and a multitude
of fungal support systems.

The Inevitable Decline: Endemic Tree Diseases

If you look around any street, urban woodlot, or nearby forest, you can always find a sick or dying tree. It takes an experienced eye to distinguish the effects of stress or drought from
microbial or insect diseases because the symptoms are often
similar. The upper branches may wilt or the leaves on a part
of the tree may fade and drop off early. Sick leaves have yellow, brown, or black spots where photosynthesis is disrupted
and the green chlorophyll fades away. Swellings or cankers on
trunks and branches might ooze sap, or the bark may crack
and expose the wood underneath. After the damage is done
and the parasite or pathogen is ready to look for a new host,
mushrooms or other spore-producing structures materialize
out of the trunk or surface roots.

There is a lot of resiliency in any individual tree. Its sophisticated defense systems flush away pests with sudden flows of
sap, infuse wood cells with toxins like tannins or oxalic acid to
inhibit microbial growth, or wall off blocks of infected tissue
with layers of dense, thick-walled cells. If an infection attacks
a single leaf, the tree drops it off. Sometimes a whole under-
performing branch is jettisoned. The tree may be weakened
a bit in the process, but it gets by. Then it signals nearby trees
to let them know trouble is in the area. Just as with human illness, most tree diseases are annoying but minor facts of life
that tend to get more serious with time. But in plantation forests, most trees don't make it to old age.

Most tree diseases are endemic, meaning they are a constant part of an ecosystem. Often, competing organisms or
the host defense systems limit the harm they cause. The root
disease caused by the honey mushroom *Armillaria* is one
example of a widespread endemic problem. Honey mushrooms are an indicator of a forest in trouble. Popping out in
rambunctious clumps like gangs of ruffians, they attack and
kill the roots, especially of trees weakened by drought, poor
soil, or air pollution. Then they stick around to rot the wood.

Most fungi in forest soils exist in a jumble of tiny colonies, but honey mushrooms can spread to occupy a huge area.
The original humongous fungus, a clone of the honey mushroom *Armillaria gallica*, was discovered near Crystal Falls,
Michigan, in the late 1980s. The colony filled a 37-acre area
(about twice the size of the grounds of the White House) and
weighed more than 21,000 pounds, a bit less than a school
bus. The mycelia of this clone had rustled through the leaf
layer of the forest soil, inch by inch, for 2,500 years. Their
tough, stringy rhizomorphs raced ahead at a rate of about

3 feet per year. After decaying roots and decimating trees, the fungus continued its expansion, linking its victims into a matrix of disease.

For the past thirty years, mycologists have engaged in friendly competitions to find bigger honey mushroom colonies. The current leader is an 8,650-year-old clone of *Armillaria ostoyae* discovered in the Malheur National Forest in Oregon in 2008. This one covers over 2,300 acres (the equivalent of 1,665 football fields) and weighs as much as 35,000 tons, or more than 200 blue whales.[13]

The patterns of these gigantic colonies are a tangle, as if they started in multiple places and were stitched together randomly like a chaotic quilt. How did that happen? The extended family of a honey mushroom can be huge—thousands or millions of identical twins may invade an area and start to grow. The colonies might have grown from spores of a single mushroom, or perhaps from several. But once they know they are part of the same clone, they merge into a magnificent three-dimensional daisy chain. The fungus remains one clone, just with more cells. In this way, they have found a mathematical loophole: one plus one can still equal one.

How could our largest life-forms remain unnoticed for so long? Are these really the Earth's biggest organisms? "Bigness" is a relative concept. It's tricky to compare living things with different body designs and different approaches to organizing themselves into units that might be considered "one." Some trees are also clones. Pando is the nickname of a trembling aspen clone in Fishlake National Forest, Utah, also known as the Trembling Giant, that includes 47,000 genetically identical trees joined together by an underground root mass. It weighs 6,615 tons (or almost 45 blue whales' worth), covers about 100 acres (about the same size as Vatican City), and may be 80,000 years old. Do we count this as one individual or 47,000?

Perhaps it is our anthropocentric definition of the word "individual" that is the problem. In our human species, we have firm ideas about what an individual is; when we meet other humans, we can be fairly certain that they are not us. Fungi and trees value their individuality less than we do, and for many, clones are an everyday part of life. They can live in independent colonies but still be part of the same clone. Maybe size is not a useful concept for organisms that aren't animals. Nevertheless, whether or not the resident clone is the biggest, it is still amazing, and Crystal Falls still celebrates its Humongous Fungus Fest each August. It features the Humongous Mushroom Pizza, baked on a 10-square-foot pan in a custom-built brick oven.

The susceptibility of a forest to disease depends on the genetic diversity of its trees. The greater the genetic variation, the higher the chance that the population will be able to adapt to new diseases or environmental changes. The trees in old-growth forests are more diverse than those in commercial plantations, which are often monocultures of one species or a single clone. For example, *Eucalyptus* (or gum tree) plantations around the world are often clones with almost no genetic diversity. If a virulent fungal disease arrives, what would be a minor illness in a mixed-growth forest might become a serious killer in a monoculture. If trees are weakened by drought, or if climate change alters the dynamics, fungi or the insects that carry them multiply exponentially and a disease becomes an epidemic. If a pathogen crosses an ocean to a land where no tree has encountered it before, even a genetically diverse population of trees may be powerless to stop it.

The Invaded Forest

My career began in the dying days of the Dutch elm disease and chestnut blight epidemics, two invasive fungal diseases that transformed rural and urban landscapes on both sides of the North Atlantic. Both elms and chestnuts were beautiful trees with tough, attractive wood used to make buildings, furniture, musical instruments, and coffins. Ornamental elms grew along urban boulevards and their uppermost branches formed leafy corridors that shaded the summer pavement. And a quarter of the trees from Georgia to Ontario were native chestnuts, enchanting giants with spreading crowns that evoked cathedral ceilings. The symptoms of the two diseases looked similar—hyphae plugged or killed the sap-carrying vascular tissue and disrupted water transport from roots to leaves. Both diseases were spread by trans-oceanic trade, but their biology was otherwise different.

Dutch elm disease (DED) was caused by two similar pathogens that passed by in separate waves. Both came from Asia. The first epidemic was noticed after World War I in the Netherlands, a discovery that gave the disease its name. Initially, the wilting was blamed on chlorine gas, the chemical weapon of choice during the war. A Dutch doctoral student, Beatrice Schwarz (1898–1969), isolated the fungus *Ophiostoma ulmi* from dying trees but her discovery was disregarded for nearly a decade.[14] After all, it was economically troublesome to delay the boatloads of lumber sailing from Europe to support the building boom in America. By the 1930s, the first wave of DED hit the U.S. and then Canada in the 1940s, spreading from eastern ports westward. A second epidemic on both sides of the Atlantic twenty or thirty years later was caused by a similar but deadlier species called *Ophiostoma novo-ulmi,* which left the Y-shaped skeletons of dead trees between farmers' fields and along urban boulevards.[15]

Both DED fungi were carried by bark beetles attracted to weakened trees. Organisms that transport pathogens are called vectors—the malaria mosquito is a well-known example. Insects often vector fungal diseases of trees, and sometimes these relationships are considered mutualistic symbioses. From the fungal viewpoint, it is more efficient to attach spores to a flying ace than to throw them randomly into the wind. Bark beetles are less than half the size of a grain of brown rice. In America, DED was vectored by a European elm bark beetle, a co-invasive that hitched along in the same cargo holds as the fungus. Upon arrival, the fungus also jumped over to a local American elm bark beetle. The moist, nutrient-rich sapwood that is the preferred food source for certain fungi also attracts the beetles. Adults drill through the bark and etch script-like tunnels along the surface of the sapwood just below the nutritious cambium of the bark. There, they lay their eggs and the larvae mature in the sheltered chambers. Slimy-headed fungal bristles line the galleries and smear spores onto any departing bark beetle that waddles past. When the new generations are ready to fly, the fungus is carried off to a new home. In infected trees, spores flow up with the sap into the youngest branches. First the tips wilt, then the leaves yellow, crumple, turn brown, and fall off. The fungus excretes a small toxic protein called cerato-ulmin, which cripples the pipe-like cells in the bark, eventually leading to wilting and death. After a year, the thick, furrowed bark loosens and falls off the trunks in huge sheets.

To try to control the epidemic, arborists initially hammered metal needles into the trunks and injected benomyl, a fungicide normally used to protect roses from powdery mildew. This treatment was expensive and needed to be re-applied every few years. Other pesticides, like chlorpyrifos, were more effective but toxic to humans. Another strategy was to exploit the pheromones released by the attacking bee-tles to attract brooding mother beetles to the weakened elms. In an attempt to scramble the chemical signals, arborists attached glue traps laced with the hormone to fence rails near elms hoping to misdirect or capture bark beetles before they attacked a weakened tree. But by the 1980s, thirty years after the start of the second epidemic, American elms were almost gone east of the Mississippi. When young elm seedlings were planted to reshade the city streets, they were quickly infected by the residual populations of *Ophiostoma* and beetles lurking in firewood.

Tender-hearted foresters at the University of Guelph har-vested seeds from the few remaining healthy elms in Canada to start the Elm Recovery Project. They called it "a dating service for lonely elms" and hoped to regenerate disease-resistant seedlings. In some cities, Chinese elms were planted. They had inherent genetic immunity to *Ophiostoma* because they came from the same area that the disease did.

Many *Ophiostoma* species are endemic in North America and Europe. Most are associated with bark beetles, some-times much more intimately than the DED fungus, but the result is a mere scab from the tree's point of view. Sometimes a minor disease results in unsightly blue staining of the wood.

Anyone flying above the ponderosa pine stands of the Pacific Northwest of North America in recent years has seen the huge swaths of once green forests that are now rusty brown. This devastation is caused by an invasion of the mountain pine beetle, *Dendroctonus ponderosae*, a native of warmer climes that no longer freezes to death in the milder twenty-first-century winters. Each beetle has a *mycangium*, a pouch designed to vector spores or hyphal bundles of symbiotic fungi, and *Dendroctonus* carries spores of a few mutualistic symbionts closely related to *Ophiostoma*—*Grosmannia clavigera* and *Leptographium longiclavatum*. After the beetles carve the tunnels, the symbionts either transform themselves into cushions of food (reminiscent of leaf-cutting ant gardens) or soften up the wood fibers for the dining pleasure of the beetles' larvae. But they can also be tree pathogens in their own right. The few pines that withstand the beetle attack are killed by the vectored fungi instead.

In 1904, twenty-eight years after Japanese chestnut seedlings were planted in New York City, bright orange cankers burst out of a native American chestnut tree in the Bronx Zoo. Chestnut blight, caused by the fungus *Cryphonectria parasitica* (Cp), enters through wounded roots, cracks in bark, or scars on broken branches. Raindrops splash its spores into the air or wash them down to the roots. The cankers girdle the trunks and strangle off the water supply. Chinese and Japanese chestnuts coped with this fungus in Asia, where it was a minor illness, but in America, the native chestnut was defenseless. Four billion trees died within forty years. Later, in several European countries, an outbreak during World War II nearly decimated the harvest of chestnuts.[16]

Foresters feared that Cp would cause the extinction of native American and European chestnut species. Surprisingly, some infected trees in Italy healed spontaneously. Healthy bark grew over and sealed the cankers. An RNA virus called CHV1 (*Cryphonectria* hypovirus 1) had infected the fungus and turned the aggressive, disease-causing blight strains into benign forms.[17] The virus-infected fungus still made small cankers but didn't kill trees anymore. The transformed fungus behaved almost as if it had turned from a pathogen into an endophyte. And when infected hyphae of Cp merged with other colonies of the same clone in nearby trees, the virus was transferred along with them. Foresters started pushing ground-up colonies of virus-infected strains into cankers on dying trees. The wounds partly healed, and the virus-infected clones migrated on their own to heal nearby trees. The epidemic slowed. Today, the Italian countryside is again filled with chestnut orchards. Some trees still have mild symptoms of Cp but not lethal infections.

Unfortunately, the virus did not spread as easily in other countries. North American orchards are afflicted by a greater variety of Cp clones, so hyphae from different colonies are less likely to merge. The virus was locked into a specific clone. To treat a forest infected by so many clones with the virus, each pathogenic clone needed a complementary virus-infected clone. Fungi apparently have contradictory requirements to link their hyphae into extended networks and to prevent the spread of viral diseases. Perhaps the genetic self-recognition system that prevents different clones from merging evolved to slow the spread of viruses.[18]

Chestnut trees take decades to mature, and during the epidemic they usually died before they made nuts, meaning it wasn't possible to grow new trees from seed. In America, tree breeders transferred some genetic disease resistance from Asian chestnut species into the native species, trying to maintain as many features of the native species as they could. Mature chestnuts returned to the United States after a fifty-year absence—but they are hybrid trees. Nature's version of the native American chestnut is essentially extinct, at least in the wild, its genetic history absorbed into a human-made hybrid. But the hope is that this new American chestnut may someday re-create an imitation of the lost Appalachian forests.

Today's rogues' gallery of invasive forest pathogens is enough to frighten any tree lover. White pine blister rust, caused by *Cronartium ribicola*, jumped from Asia to Europe to North America about a hundred years ago and is now watching me from its perch in a shabby-looking tree outside my window. Ash dieback is caused by *Hymenoscyphus fraxineus*, which probably rode along as an endophyte in seedlings shipped to Poland from Asia. In the 1990s, it jumped into native ash where it was a pathogen instead, and it is now wiping out its new host through much of northern Europe. Beetle-transmitted fungi, like the colorfully named thousand cankers disease caused by *Geosmithia morbida*, celebrated the millennium by attacking black walnut trees in the United States; its origin is unknown.[19]

Breaking It All Down: Decay and Nutrient Cycling

Whether from disease, aging, drought, or forest fires, plant matter constantly ends up on the ground. The debris is crammed full of structural and storage carbohydrates like cellulose, lignin, and starch, as well as stubborn secondary metabolites like tannins and melanins—all of which need specific enzymes to break them down. The capture of carbon molecules from the air by photosynthesis, the assembly and consumption of energy-rich organic storage compounds by living organisms, and the subsequent breakdown of these compounds and release of carbon dioxide gas back to the atmosphere is called the carbon cycle. The amount of carbon in the world is constant, but the ratio between what is in the air and what is incorporated into minerals or living cells varies. A lot was locked up and kept out of circulation in the form of so-called ancient carbon, which was part of living cells in prehistoric times but was sequestered for millions of years in the deposits we now burn as fossil fuels.

Eighty percent of the world's actively cycling carbon is locked up in the cellulose and lignin in the cell walls of plants.[20] Cellulose, the molecule that makes up paper and cellophane, is composed of chains of glucose molecules bundled into fibers. Lignin is an irregular three-dimensional matrix of phenylpropane alcohols that glues the cellulose fibers together and makes wood hard. Fungi, especially ascos and basidios, are the main organisms that free carbon bound in plant cells and wood, using enzymes like cellulase and ligninase to generate energy for themselves and ultimately the other kingdoms of life. Without fungal decomposition, plant debris would pile up endlessly and bury us all.

Decayed wood is often hidden in the core of living trees. We are usually surprised when a severe storm causes a seemingly healthy street tree to drop heavy branches or cracks the outer cylinder of sapwood, tipping a splintered mass of trunk and boughs onto houses or cars. Rot often begins when spores germinate in wounds left by broken branches or in bark scars gouged out by falling trees. You can learn to recognize its signs, like subtle changes in color, dead branches, cracks or wounds in bark, or dramatic cavities hollowed out by woodpeckers. Bracket fungi or mushrooms pushing out through bark are clear clues that the wood is no longer as solid as it was.

When an old tree falls or is cut down, the consequences of hidden decay are exposed. The hollow center is called "heart rot." The cavity forms when growing mycelium bypasses the bark and sapwood and bores straight through to the core of the trunk, where the wood is drier and cellular defense systems are less active. The tree loses its central support and can't endure buffeting winds or heavy snow. Heartwood is home to a particular assortment of fungi. For example, *Ganoderma applanatum* causes dramatic decay in trunks of broad-leaved trees near the ground, a phenomenon known as butt rot. Folk artists scratch sketches into the white pore surfaces beneath its large half-moon-shaped red caps, giving it the common name "artist's conk." Its close relative *Ganoderma lucidum*, the reishi mushroom, makes antler- or coral-like growths in low carbon dioxide conditions and is known for its alleged anticancer properties. Many of these wood-decaying fungi are considered pathogens. Their activities are confined to the nonliving part of the tree, the heartwood, but the living part of the tree may lack the strength to keep itself upright, leading to an inevitable catastrophe.

Rotting wood gets soft enough that you can pick it apart with your fingers, and sometimes you can squeeze water from a handful of fibers. That water is released from cellulose and lignin molecules that are broken apart during the decay. Carbon dioxide gas drifts out as a by-product. Several basidios, especially mushrooms and polypores, and microscopic asco moulds make cellulase enzymes that nick the long sugar chains of cellulose into shorter chainlets. As cellulose breaks down, the wood gradually loses structure. It eventually leaves behind a shrunken, crumbly residue called brown rot that is mostly lignin. Conifer wood is most prone to brown rot. In addition to attacking cellulose, some basidios make ligninase enzymes that crack the honeycomb of lignin molecules into fragments, leaving behind wood that looks bleached. This pattern of decay, more common in the wood of broad-leaved trees, is called white rot.[21]

The predictable sequence of species that occurs during the breakdown of organic matter is a type of ecological *succession*.

Waves of different fungi, microbes, and micro-animals attack organic matter. As species process their favorite foods, they break complex molecules into simpler ones and render the result more digestible for a later cohort. In trees, the original basidio pathogens begin the decay of the cellulose and lignin in the heartwood or sapwood of the living plant. Once the wood is exposed to the air, spores of saprobic basidios, like oyster mushrooms (*Pleurotus* species), land and continue the decay. The original pathogen fades away. As more polysaccharides break down, the wood fiber becomes wetter and sweeter, and a diversity of saprobic moulds moves in, often so aggressive that they kill off the wood-decay fungi.

The soil around trees is an active compost. Decaying wood on the forest floor mixes with the remains of leaves that flutter down from tree crowns, and bark chips that slough off expanding tree trunks. Work crews of fungal sanitary engineers—mostly free-living asco and zygo moulds that are never found anywhere else—penetrate the dead plant material. Endophytes land and switch over to saprobic feeding, leaving behind the dead tissues of their former leaf homes as honeycombed skeletons. Rhizosphere fungi are saprobes that hover near roots without connecting to them, scavenging and releasing nutrients. The microfungal diversity in and on decaying leaf litter and in soil is immense, especially in tropical forests where hundreds of moulds attack the shed leaves. A multispecies soup of enzymes attacks the chemical bonds of the unwieldy carbohydrates and breaks the organic matter into smaller fragments. Bacteria nudge in and add their enzymes, then the detritus is gobbled up and ground down further in the digestive tracts of insects, worms, and other tiny soil animals.

The output is the soil-like organic compost known as humus. In temperate forests debris keeps falling from above, and the soil settles into visible layers (horizons) that are marked by steps in the breakdown from freshly fallen tissue to humus. Particular fungi favor particular horizons of soil, and the profile reflects the succession of microbes that unfolds during biodegradation. Humus absorbs water and oxygen and provides easily digested nutrients, acting almost like a dietary fiber to keep the forest healthy. Seeds bed down in this fertile layer, moulds soften their hard outer coats, and then they germinate, sending new shoots up into the light.

Biological Control: Applied Symbiosis

With endophytes running through branches and leaf tissue, epiphytes covering foliage, rhizosphere fungi surrounding roots, mycorrhizae connecting tree to tree in the soil, and saprobes processing the fallen debris on the ground, a healthy forest is a busy place. Keeping forests healthy is a long-term commitment. Forest ecologists and forest pathologists (sometimes called tree doctors) try to stay ahead of diseases that appear in the woodlots in their care, and treat or remove symptomatic trees before infections spread. Spruce budworm outbreaks, for example, occur in roughly thirty-year cycles in several parts of North America—the timing is not well understood, so careful monitoring is needed. Then, unless the area is bombed with chemical insecticides from a low-flying aircraft, entire forests can be destroyed as the voracious moth larvae devour needles or leaves.

In afflicted forests, however, some stubborn old trees do make it through multiple budworm epidemics. These trees should have the best endophytes, which might make high concentrations of mycotoxins to deter hungry larvae. Foresters have wondered if replacing the random mixture of endophytes in susceptible trees with carefully selected strains that produce lots of toxin might be a better way to protect the forest.

To look for such fungi, my colleagues and I traveled with some local foresters and a mysteriously taciturn man into the remote corners of the Acadian forest of eastern Canada. The four-wheel-drive jostled on heavily eroded tracks through cleared plantations, but the scouts knew the precise coordinates of the elite spruce—the tallest ones, with the straightest trunks and healthiest needles. The stranger in the SUV was a sharpshooter with a high-powered rifle. After scanning the tall trees through binoculars, the chemist in the group, who was leading the effort, pointed and said, "That one." The sniper shot, and a branch tumbled from the treetops through the foliage to the ground.[22]

Back in the lab, students and biologists went through the long process of isolating and purifying hundreds of endophyte cultures from the branches we'd collected and other specimens culled from nearby trees. With DNA sequencing, we recognized that many of the crinkled brown cultures that crawled out of the needles were *Lophodermium*, even though they made no spores. Several species of another asco called *Phialocephala* crept out of other samples; sometimes, after several months in the refrigerator, they made a few spores. Cultures of both fungi made mycotoxins that inhibited or killed spruce budworm larvae in lab experiments, and others that slowed the growth of other fungi. We hoped that these endophytes would protect the trees from the caterpillars in the forest. The questions were how to get them into living seedlings and whether they would take up permanent residence as endophytes and reduce the losses expected during budworm epidemics. Early experiments with young trees were encouraging. The endophytes established in young endophyte-free seedlings and survived for several years. The same protective mycotoxins showed up in the needles.

Today, many conifers begin life in nurseries, gymnasium-sized greenhouses that shelter millions of seedlings in custom-designed containers that look like deep egg cartons. The seedlings are lined up on rows of tables under banks of fluorescent lights, with the hoses and pipes of the irrigation system running above them. The endophytes grow better in liquid broth in fermentation tanks than they do on agar, so the blobs of mycelium from the Petri dishes are blended into a slurry. To introduce the fungus into the seedlings, hyphal fragments are sprayed from the overhead irrigation system when the young needles are most receptive to colonization. Then the seedlings are planted out into the forest and the endophytes persist, penetrating new bud scars and colonizing needles as they appear each spring. The anti-insect metabolites build up inside the needles where they are needed. When the old needles fall to the ground and decay, the mycotoxins also break down. Hundreds of millions of spruce seedlings have now been treated with carefully selected endophytes native to eastern Canada. This is a long-term experiment; the people who set it up may not be working anymore when the trees mature and the next epidemic arrives.[23]

The deliberate use of endophytes to reduce insect outbreaks is an unusual example of biocontrol but an instructive one. First, the treatment applied to seedlings is a much more focused strategy than spraying hundreds of square miles in a forest. It redirects or amplifies an existing mutualistic symbiosis in favor of desirable strains that produce potent mycotoxins. It is an intervention to ensure that planted seedlings will have an effective, protective endophyte population instead of a haphazard mixture of less helpful strains.

Second, this approach suggests that if we want to manage ecology for our own benefit, as we try to do in both wild and planted forests, we need to be aware of all the organisms that are present. The collection of microbes on the surface of and within the body of a host is called a *microbiome*. When the host is a living plant, the associated microbes are called the *phytobiome*. Trees are multispecies conglomerates called *holobionts*. To our eyes they seem to be individuals made up of cells of one species. But when examined more carefully, we see each one is assembled from many different component species. To improve how a managed forest works, we need to understand how tree phytobiomes form, how each component species changes over time and adapts when threatened, and how all the components cooperate to create something we regard as a tree. Then, if necessary, we can tweak the holobiont to optimize its health and resilience. After all, the holobiont that we see as a tree is as much about fungi—the endophytes, mycorrhizae, epiphytes, rhizosphere microbes, and the pathogens—as it is about the plant that holds the pieces together. Simply put, tree huggers will always be fungus huggers—through all phases of a tree's life.

When we think about trees and fungi, we tend to think only of forests. But whether the trees are naturally occurring or planted—in forests, plantations, or orchards—fungi are an integral part of the phytobiome. And trees and forests are as important for producing mushrooms, spices, and fruit as they are for producing lumber. The phytobiome is also important on farms, where understanding the beneficial and harmful influences of fungi may make the difference between producing enough food for our growing population—or not.

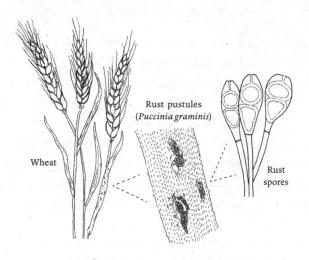

Wheat

Rust pustules
(*Puccinia graminis*)

Rust
spores

4 | *FARMING*

The Seventh-Oldest Profession

AGRICULTURE WAS the first human enterprise we might call biotechnology—the deliberate adoption of the biological or biochemical talents of other living things for our own benefit.[1] Like most great ideas, it was reinvented repeatedly, perhaps ten times by different human tribes around the globe. Before agriculture, there were fewer than fifteen million humans; now we are more than seven billion. We may be nostalgic about our hunter-gatherer origins, or long for a nomadic foraging lifestyle and Paleo diet, but those idyllic days are gone.[2]

Today's grain farms are a completely different world from a natural meadow or prairie, though above the surface they seem similar. They are artificial ecologies created over centuries, as farmers tilled their soil, planted seeds, prayed for sun and rain, and agonized over failed crops long before the causes were understood. They tinkered with their domesticated crops, breeding plants with desirable characteristics without realizing that symbionts were involved. In recent times, in the rush to produce ever more food, high-volume, high-throughput agro-industrial factory-like farms have been assembled without a full understanding of all the moving parts. Domesticated seeds and livestock, fertilizers and pesticides, threshers and combines, and irrigation systems are all hallmarks of modern farming. But the symbioses between crop plants, fungi, and other essential microbes that occur in wild plants and uncultivated soil were not considered important until recently.

Most contemporary farmers have a strong understanding of biology, and they continue a long tradition of planting enough to compensate for an expected 10 to 20 percent loss. Striving to provide missing nutrients and drive off diseases and pests, generations of farmers have relied on synthetic chemicals to tweak their crops, and many growers still use them. Plants often have trouble absorbing phosphorus and potassium from the soil. And although nitrogen gas makes up 78 percent of the atmosphere, plants, animals, and fungi cannot absorb this essential element from the air by themselves, even though they need ammonium, nitrate, or nitrite ions to assemble into amino acids and then proteins. Only nitrogen-fixing bacteria can make use of the gas for themselves.[3] Luckily, many of them are mutualistic symbionts that form swollen nodules on legume roots, live free in soil or water, or partner with other organisms to pass their nitrogen on. In the early part of the twentieth century, German chemists invented a process to scrub nitrogen from the air so that it could be combined into fertilizers designed for crops that suffer nitrogen deficiencies. The resulting increase in food yields, without a need for more acres of farmland, was a main factor in the huge population growth of the past century.

It's easy to understand why microscopic partners are so often overlooked in farming, which focuses on the big scale. But phytobiomes of crop plants are as important as the phytobiomes of trees. Many moulds affect plant health, and we need to understand them if we want to maximize the efficiency of our agricultural factories. Restoring the function of these natural biological modules of agricultural plants may be the key to increasing productivity while minimizing the environmental damage attributed to modern agriculture.

Fungal Crop Symbionts: Mycorrhizae, Rhizosphere Fungi, and Endophytes

There are pictures of me as a four-year-old on my uncle's farm, posing with a bundle of freshly threshed wheat clutched to my chest. Holidays for our family meant driving the 1,300 miles across Ontario and Manitoba in our 1960 green Oldsmobile until we pulled in at the family farmstead north of Regina, Saskatchewan. I always loved the prairies. My uncle Hugo was an old-fashioned man with pomade in his center-parted hair, pince-nez glasses, and a Bakelite cigarette holder. Lush fields of wheat drove his imagination. Each evening before sunset, we'd saunter out to inspect a different quarter section of the farm. The wheat was about the same height as me and I could almost disappear into it. To my eyes it seemed like a healthy crop, but if I'd known to look, I would have realized that something was amiss. The grasses and weeds growing on the margins of the field were filled with symbiotic fungi, but there were far fewer of them in the crop plants themselves.

Compared with trees, where the vibrant symbiotic relationships involving fungi drive the forest, the symbiotic relationships on farms seem feeble. It's difficult to compare a farm ecosystem with a balanced community in nature because, in the process of domestication, we've unwittingly disrupted many of the original biological connections. The fungi that are symbiotic with crop plants are found mostly around their roots. If you sieve the dirt from beneath the plants in almost any field, you'll find hundreds of shiny brown balls, each smaller than the tip of a pin. Your first guess might be that they are tiny insect eggs, but they are giant single-celled asexual spores. When these spores are stained and viewed through a microscope that casts ultraviolet light, you can see hundreds or thousands of glowing nuclei.[4] We aren't quite sure why they have so many, but we know individual spores are connected by a thin weft of hyphae (which also have lots of nuclei in their cells) to specialized structures inside the roots. Hence they are often called *endomycorrhizae*, which means "inside the roots."

Endomycorrhizal fungi evolved a couple of times, once in the phylum Glomeromycota and once as part of a group of zygos known as the Endogonomycetes (now included in the phylum Mucoromycota). When you look at thinly sliced root tissues under a microscope, fungi of both groups make one or two distinctive structures inside individual cells: round, swollen cells called vesicles and elaborately branched tree-like structures called arbuscules. The latter give these symbioses the name arbuscular mycorrhizae (AM). Vesicles of AM fungi were probably illustrated by the Swiss botanist Carl Nägeli (1817–1891) forty years before the Hartig nets of ectomycorrhizae were discovered, but nobody understood what they were. They look like structures made by common plant pathogens called *Pythium* (fungus-like relatives of the potato blight), and for more than a hundred years they were thought to belong to a rare kind of parasite. The debate about whether they might be beneficial started in the 1950s. Then in the 1970s new microscopic techniques for examining roots were introduced, and we realized that about 70 percent of wild plants have AM roots, including tropical and grassland plants, as well as some tree groups that don't have ectomycorrhizae.[5] Arbuscules and vesicles are both well enough protected inside roots to be preserved in root fossils, but it took a while for paleontologists to realize that they were part of a fungus. This fossil record, combined with evolutionary dating using molecular clocks, suggests that AM symbioses already existed when plants migrated from oceans onto land. Without this support, it seems unlikely that land plants would be the dominant life-forms they are today. AM researchers are fond of claiming that "plants don't have roots, they have mycorrhizae," suggesting that the plants' own roots are a plan B in case AM fungi aren't around.

Unlike most ectomycorrhizal fungi, AM fungi have a promiscuous, open-arms policy towards their plant partners and form symbioses with many different plants. Thousands of plant species rely on only two or three hundred species of AM fungi. As with ectomycorrhizae, hyphae of AM fungi meander through the soil searching out deposits rich in phosphorus and other minerals. The nutrients stream back through the hyphal plumbing system and are dispersed through the finest branches of the arbuscules directly into the root cells. The repeated branching of the arbuscules, which look something like the alveoli in our lungs, maximizes the surface area to exchange minerals and liquids for photosynthetic carbon from the host plant. Some but not all AM fungi make vesicles to store a backup supply of nutrients for hard times, like a pantry for the symbiosis.

Plant breeders historically did not pay attention to AM fungi, and the ability of new cultivars to form endomycor- rhizae was not considered. Despite this oversight, now that we are aware of them, we see that AM fungi persist in farm soils from one year to the next and that mycorrhizae still form in many crop plants. But the diversity of fungi involved is lower and the increase in root mass is less pronounced than in nearby uncultivated soils. Nevertheless, even plants casu- ally colonized by AM fungi absorb more minerals and gen- erate more photosynthetic carbon and hormones than those without. The fungi can also bind excess heavy metals into the chitin in their cell walls and to structures inside their cells, reducing the toxic effect of these ions on the plant. The result is greener, denser foliage and more seeds. Probably because of their more extensive root systems, mycorrhizal plants are better at tolerating drought. AM fungi seem to prime plant defense systems and reduce infections by root pathogens, but we do not know much about how this process works. Since about 2000, evidence has been growing that mycelia of AM fungi can form common mycorrhizal networks among plants of the same or different species, analogous to the Wood Wide Web in forests. The connections seem to provide dawdling seedlings with a boost of nutrients, share the priming signals alerting them to the presence of plant pathogens, and move signaling hormones from one plant to another. But seasonal plowing of fields disrupts these networks.[6]

With so many benefits from AM, you might expect that adding the fungi to fields would be a no-brainer. Unfortu- nately, it is not so simple. Positive effects in greenhouse experi- ments, where we can compare plants grown in sterilized soil with and without mycorrhizae, often aren't matched in field experiments where there are always AM fungi waiting to colonize plants. The relationships between endomycorrhizae and plowing, fertilizer and pesticide use, plant diseases and crop yields are not straightforward. The nutrient-scavenging functions that the fungi provide for the plants are made redundant by the abundant nutrients provided by synthetic fertilizers. Most cultivars grown today were bred to expect high amounts of nutrients from fertilizers and are not attuned to working with the lower levels provided by mycorrhizae. Fungicides intended to kill plant pathogens may also kill symbionts. Further, although reduced plowing usually increases AM diversity, some species have difficulty colonizing plants in untilled fields, perhaps because of increased competition from saprobic fungi growing in the surface layers of crop debris. As a result, adding AM fungi often shows little benefit in crops.

The majority of AM species don't grow at all in the usual agar systems used to culture most other fungi. Their hyphae need to be near living roots before they will grow and make more spores. To culture these species, plant biologists either grow root cells in agar culture and then colonize them with spores picked out of soil, or they inoculate spores of promiscuous AM strains into pots with fast-growing mother plants. After a few months, the cyst-like spores can be sifted out of the medium and mixed with powders and granules that keep them alive long enough for you to buy them at your local garden shop. Agricultural companies have developed industrial-scale methods to grow AM inoculum for fields that cover hundreds of acres. Usually, the fungi *Rhizophagus intraradices*, *R. irregularis*, and *Funneliformis mosseae* are used either alone or more often in mixtures. Such microbial inoculants, sometimes called biofertilizers, may be the way of the future, but we have a lot to learn about how AM fungi work. It will be some time until we can maximize their effects, and there are probably a few decades of plant breeding ahead to develop cultivars that will give reliable high yields when grown with blends of high-performing (or elite) AM fungi.

Other fungi also live near the roots of crop plants but do not form mycorrhizae. DNA surveys detect thousands of these non-mycorrhizal rhizosphere fungi and bacteria mooching around in grasslands, but only a subset lives in agricultural fields. Soil fertility improves as fungal diversity increases. The saprobic rhizosphere fungi that do live in fields appreciate the mixing of soil and roots during plowing, and their hyphae swarm the straw and dead bits of leaves and roots. There they continue the endless job of breaking down carbon and releasing other minerals, loosening soil particles to allow water and air flow, assaulting pest insects or nematodes, and attacking or outcompeting plant pathogens. Some of the rhizosphere moulds, like *Penicillium bilaiae* and *Trichoderma virens*, grow easily in culture and are being developed as commercial products to boost plant growth. Their spores are mixed into powders that are sprinkled into garden or farm soils during seeding or transplanting. When their hyphae start to grow, they exude organic acids that convert minerals into ions easily absorbed by plants. They don't seem to interfere with AM fungi and can be helpful biofertilizers in low-phosphorus soils.[7]

In addition to the mutualistic fungi around their roots, most wild grasses have endophytes in their leaves. Because most grass plants live less than a year, these symbionts have little time to set up house. So, unlike their counterparts in the forest, grass endophytes spend their whole lives inside plants and then are prepackaged in the seeds. When the grass plant withers away and releases its seeds on their journey, the endophytes ride along, already installed when the next generation of seedlings starts growing. *Epichloë* species are mutualistic endophytes in many species of wild grass. Their growth outside the plant is so sparse you can hardly find them with a microscope, but inside stems they form a meshy network that produces complex toxins called alkaloids. These compounds protect the tissues from insects but not always from other animals. The well-known hallucinogenic properties of morning glory seeds result from ergot alkaloids produced by endophytic symbionts (more on ergot to come).

In crop plants, some endophytes have a kind of Jekyll and Hyde personality. As we've seen in forests, they're useful to plants because they ward off harmful insects. However, if sheep or ponies consume too many endophyte-laden leaves, in particular the forage grass fescue, they lurch around and fall over like drunken college students, a malady known as ryegrass staggers. If they eat a particularly heavy dose, they develop a condition called fescue foot and their hooves can fall off. The toxins don't affect birds or rodents that snack on colonized seeds, but wise rabbits turn their noses up when offered pellets made from endophyte-colonized grass.

The same symbiosis is beneficial in some situations and detrimental in others, depending on whether pest insects or livestock are consuming the plant. Proactive farmers grow endophyte-free forage for their livestock to prevent illness and keep the endophyte-plus seeds for their lawns.[8] Some endophytes also turn against the same plants they at first seem to help. If you wander through a meadow in the last half of summer, you often find grasses with stems sheathed by the telltale speckled ascospore-producing tissue of *Epichloë*. The coating looks a bit like egg yolk and is the first symptom of a disease commonly known as "choke." At this stage, near the end of the grass plant's life, the fungus switches from mutualist to parasite and squeezes off the nutrients flowing from the roots into the upper part of the plant.

Endophytes are actually rare in the cultivated grains we grow for food. The grains sown on modern farms were domesticated from wild grass plants thousands of years ago, starting when Neolithic farmers saved their favorite seeds. The ancestral plants probably had endophytes, but they disappeared during the breeding process. Some botanists have tried to adapt *Epichloë* species from wild plants to grow in grain plants, in the hope that the endophyte will protect against insect pests. It works to some extent with wheat in greenhouses, but unlike the biocontrol system developed for conifer needles, so far no stable long-term symbioses have been established. For the time being most of our grain crops will continue to grow as endophyte-free plantations, but eventually this may change.

We may have domesticated many grains and grasses by selective breeding for our own purposes, but many of them have also done very well through their association with us. Wheat was just a minor wild grass from the Middle East before it got involved with humans ten thousand years ago.[9] Now it occupies about 1.5 million square miles of farmland around the globe, including hundreds of varieties bred to improve yields. However, cosmopolitan crops tend to have cosmopolitan diseases, following along as acreages expand, seeds are distributed from one country to another, and food is imported and exported.

Rust Never Sleeps

My grandfather grew Marquis (prairie farmers always called it "Mark-wiss"), a high-yielding wheat variety that matured quickly in the short prairie summers but was susceptible to black stem rust. The farmhands often emerged from their fields with their shoes and pants puffing orange clouds of spores. The combines that separated the wheat from the chaff caked up with the same powder. Stripes of tangerine-colored pustules on wheat stems were caused by the rust fungus *Puccinia graminis*. There are thousands of rust species, most pathogens of specific plants, including most grains and many trees. Each blister fills with microscopic spores. A moderately infected wheat field might have 50 trillion stem rust spores per acre.

Known since biblical times (Genesis 41:25–30), black stem rust was first noticed in the grain belts of Canada and the United States in the 1870s, then gradually spread as it increased in virulence. Wheat rust epidemics became increasingly severe on the North American prairies after the turn of the twentieth century. Before the United States entered World War I in 1917, a rust epidemic wiped out a third of the harvest and bread prices skyrocketed. To cripple the fungus, the first step was to stop it from having sex, which meant searching out and destroying its alternate host: barberry. European immigrants had brought ornamental barberry shrubs to the Americas, a source of berries for making preserves. Boy Scouts and "Rustbuster" clubs were soon mobilized with instructions from the Rust Prevention Association and the U.S. Department of Agriculture to "execute this criminal bush wherever it is," because it was pro-German.[10] The obliteration of the alternate host slowed the evolution of new pathogenic strains, and for a while, rust outbreaks were less severe. The 1930s, however, was a decade of drought and economic and political turmoil in Canada and the United States. Any crop that did grow was felled by rust. As the Great Depression took hold, a generation learned the hard way that 'hoppers or a fungus could kill a farm.

When rust spores land on wheat, they germinate and specialized hyphae punch their way through the hard cell walls of the stem into the softer living tissues. The mycelium then meanders through the plant, out of view, reminiscent of an endophyte, but there is no beneficial symbiosis here. The disease either kills the plant or siphons off so many nutrients that the seeds can't mature. The complexity of the life cycle of wheat rust was discovered by 1660 and rivals some insect metamorphoses. Five different spore types—four asexual and one sexual—form on two different, often distantly related, host plants (known as alternate hosts) in different seasons. Some of these spores spread the contagion for a short distance in one field. One type develops thick walls and stays where it is, hunkering down for survival. Others float upwards for long-term dispersal on the wind. To this day, students shudder when asked to recite the details, but this complexity also confounds attempts by farmers to control the disease.[11]

Pustules of *Puccinia*'s sexual spores form on the underside of leaves, but any spores shed by infections in Canada and the northern United States are killed each winter by the cold. In 1935, when the National Geographic Society released the *Explorer II* high-altitude weather balloon, it found the distinctively spiny, golden spores of rusts in the stratosphere, 13.7 miles overhead.[12] Wheat rust uses wild barberry bushes in the tropical and subtropical Americas as a winter resort. Thermal air currents lift the spores from these plants into the northbound jet stream. Weeks later they land on tender emerging stems and leaves of North American prairie wheat. Stem rust is an invasive disease that just keeps on invading.

Epidemics of wheat rust brought a new urgency to plant breeding. Even before scientists knew how genes operated, it seemed clear that wheat and rust were coevolving. Pathogenic strains of rust are called races, and each race has a unique collection of pathogenicity genes. Farmers therefore planted seeds bred with resistance genes designed to protect them from the races occurring in their regions. Of course, the rust is constantly evolving new variants, trying to find new ways in. But until it does, the crops might have ten or twenty years free from the disease. Eventually the fungus stumbles across new pathogenicity genes that let it attack wheat cultivars that were previously immune. Plant breeders seek out new resistance genes from related wild plants and cross them into new wheat lines so that another cultivar is ready when the old one crashes.[13] When a planted crop is resistant to the local rust races, few fungicides are needed. But once a new pathogenic race finds a susceptible plant, the various asexual parts of the life cycle amplify it exponentially. A new epidemic spreads, and the genetic arms race continues.

Norman Borlaug (1914–2009) was an American plant breeder who developed wheat cultivars for the developing world. In 1944, the Rockefeller Foundation assigned him to solve the problem of wheat rust epidemics in Mexico, which were leading to serious food shortages. After World War II, India and the newly separated Pakistan suffered devastating epidemics of wheat rust that led to terrible food shortages there as well. Borlaug's colleague, M. S. Swaminathan, convinced the reluctant Indian government to plant the varieties developed for Mexico. The Indian authorities placed a massive order for an immediate supply. Borlaug shepherded truckloads of seeds from Mexican nurseries over the border into the United States. They were loaded onto barges in Los Angeles during the Watts race riots and docked in India just in time to be planted. Between 1965 and 1970, wheat yields doubled on the subcontinent. Borlaug and Swaminathan's efforts saved as many as a billion lives. Borlaug was awarded the Nobel Peace Prize in 1970. Although his efforts prevented famine, his award was controversial and remains so today. The breeding of crops for physical traits and high yield, together with the use of extensive irrigation, fertilizers, and pesticides, was branded the Green Revolution. Critics questioned the long-term sustainability of such a high-tech approach to solving problems of the developing world. Advocates of population control in the developing world were also uncomfortable. Pacifists noted that much of the fertilizer was produced by chemical companies also heavily invested in explosives.[14]

In 1998, a virulent new race of wheat rust called Ug99 was found in Uganda. No resistance genes were known that would stop it. It survives at higher altitudes than other races, which put critical agricultural regions at risk if it were to spread. After its discovery, it flew from East Africa to Yemen and Iran on the trans-Asian jet stream. Wheat breeders in Canada and the United States began searching for new resistance genes.[15] They sowed spores of Ug99 onto experimental varieties in high-security greenhouses, and in Canada, as an extra precaution, the experiments were done only in winter. It is working; new resistance genes are ready to be deployed, though new variants of Ug99 are still being discovered. So far, this rust has not touched India and remains far from the breadbasket of the North American prairies.

Poison by Degrees: Ergotism and Mycotoxins

Ergot was more of a curiosity in Saskatchewan in my uncle's time, but everyone knew the lurid backstory. During the Dark Ages, ergotism caused violent convulsions in European peasants, who also hallucinated that their limbs were aflame. In severe cases, their circulation was badly impaired and they lost limbs to gangrene. This illness came to be known as St. Anthony's Fire, named for the patron saint of the order of monks who cared for the afflicted. By 1670 the symptoms were linked to rye, a staple food of the poor, but the connection to the ergot fungus was not made until the 1880s. Two Frenchmen, the physician Charles Tulasne (1814–1884) and his lawyer brother Louis-René (1815–1885), connected several different spore forms to the ergot disease, which they called *Claviceps purpurea.*[16]

Rye crops suffer the worst ergot infections, but most wild grasses and agricultural grain plants get the disease. The fungus shoots ascospores into the spring winds just when grasses and grains are in flower—this is a common trick with plant pathogens, making sexual spores when the host plant is most vulnerable. The ergot spores germinate and send hyphae down the central channel of the young floret, taking advantage of a pathway intended by the plant to guide pollen. Each ovary is replaced by a black claw (a hard mass of mycelium called a *sclerotium*), and then no seed can form. The sclerotium gives the fungus its common name: in French, *ergot* is the hooked, backward-facing talon of a chicken, which is just what these spikes look like protruding from the grain. A sticky yellowish liquid called honeydew drips out of the florets and flows down the stalk, filled with millions of tiny asexual spores that stick to the legs of curious insects. The bugs fly from plant to plant, casually dropping off spores that start new infections. After the infected crop matures, the sclerotia fall to the ground or are harvested along with the grain, where most are sifted out in the elevators. In the spring, having survived the winter, tiny sexual structures that look like orange light bulbs balanced on curved stalks sprout from the buried sclerotia, releasing spores and starting the cycle again.[17]

Ergot is more or less a nuisance to the plant, but serious consequences await mammals who eat the sclerotia. Spontaneous abortions often occur in pregnant livestock who eat leftover grain contaminated with ergot, especially pigs. Alkaloid toxins are found in the sclerotia, especially high amounts of ergotamine, which the fungus makes to discourage insects from devouring them. In humans, ergotamine causes blood vessels to constrict. Doctors use small amounts to treat migraines, but larger doses lead to the painful burning sensations and hallucinations that are the hallmarks of St. Anthony's Fire.[18] Meanwhile, ergot can still be a significant problem in some grain belts. The sclerotia can be sieved out but are so potent they have to be handled as toxic waste.

In the 1940s, the Swiss chemist Albert Hofmann (1906–2008) was studying ergotamine as a possible drug to ease difficult human pregnancies. To improve its medical properties, he was experimenting with chemical alterations and synthesized a molecule that he called lysergic acid diethylamide (LSD). On April 16, 1943, some concentrated LSD splashed onto his skin just as he set off for home on his bicycle. He became disoriented and the sky spun like a kaleidoscope as he weaved all over the roadway trying to avoid other vehicles. Alarmed by the molecule's potency, Hofmann next deliberately swallowed a quarter gram, which would today be considered a massive dose. It was the first Electric Kool-Aid Acid Test. Despite histories of illicit recreational experimentation, LSD and psychoactive metabolites like psi-locybin from "magic mushrooms" (*Psilocybe* species) are in experimental use to treat alcoholism and depression, study chemical signals in our brains, and stimulate creativity, but at about one-tenth of the hallucinogenic dose. Psychedelia aside, the powerful effects of ergotamine and related compounds show that fungal metabolites, or chemical alterations of them, can have astonishing physiological and psychological effects.[19]

Ergot alkaloids, which include the very similar compounds produced by grass endophytes like *Epichloë*, are examples of fungal metabolites called mycotoxins. These are natural compounds that have negative effects on animals or people who consume contaminated food. In nature, mycotoxins are released by moulds when they grow in seeds or leaves and are probably intended to deter insects. Several are produced on living crops in the field and remain in the crop after harvest. We've known about them for less than a hundred years, and most people are unaware of them. But they are among the most serious dietary problems facing us today. According to the Food and Agriculture Organization of the United Nations (FAO), a quarter of our crops are contaminated with unacceptable levels of mycotoxins. Overexposure is hugely detrimental to public health, and one of the main factors separating rich and impoverished nations. Reduced pesticide use, open-air crop storage, homegrown seeds, and lax government regulation of food safety all correlate with higher mycotoxin intake.[20]

Aflatoxin, Vomitoxin, and Modern Agriculture
After World War II, high-quality protein was in short supply in the United Kingdom and food rationing continued for

years. In the 1960s, about 100,000 young turkeys were fed peanut meal made from nuts imported from Brazil that were too scuzzy to be used as food for humans. The birds convulsed—their necks spasmed and jerked and left them staring straight up—until they fell into a coma and died. The newspapers called it Turkey X disease. Unraveling the cause led to the discovery of one of the most toxic natural compounds known, a mycotoxin called aflatoxin that was rife in peanut butter but also traced a messy line back to maize.

The name "aflatoxin" comes from the first syllables of the two parts of the scientific name of the mould that produces it, *Aspergillus flavus*. This beautiful fungus—which produces a long, slender stalk with a dusty head of yellowish-green spores—overgrows nuts, legumes, cereal grains, and soil almost everywhere, but especially in warm climates. In Africa and poorer parts of Asia, the chartreuse powder of *A. flavus* is pervasive. In nature, the toxin deters insects, and perhaps birds, interested in those same seeds. When aflatoxin was discovered in the aftermath of Turkey X disease, awareness slowly grew that it was a significant contributor to human illness too. Where peanuts were a staple protein source, as in much of Africa, aflatoxin was an ancient, but overlooked, problem. An already dire situation worsened with the introduction of maize during the Green Revolution, because it provided the local *A. flavus* with a rich new food source.

Widespread, chronic aflatoxin poisoning was then found in humans. A lifetime of consuming low levels of aflatoxin in maize and peanuts, or foods including them as ingredients, catches up to us as it accumulates in the liver. If it reaches a critical level, the toxin causes hemorrhaging and cirrhosis, and it is one of the leading causes of liver cancer. Developed countries strictly regulate it in imported food, but it is hardly monitored in countries with the highest levels. Tropical countries send their cleanest crops for export because they need foreign cash. The remainder, too toxic to be exported, is fed to animals. And in countries facing malnutrition or famine, contaminated food is used, mostly unwittingly, for people. In parts of Africa, you can buy dog food that is labeled aflatoxin-free, but there are no guarantees for the food you give to your children. Daily allotments of peanut butter, which some governments provide to disadvantaged schoolchildren, are often prepared from cheap peanuts laden with aflatoxin. According to some estimates, aflatoxin kills more people than malaria.[21]

Aflatoxin is not just an African problem. Losses to the United States maize crop each year vary from $52 million to $1.7 billion. And if corn products are contaminated— like cornstarch, a nearly ubiquitous ingredient in processed foods—the toxin remains even after cooking. In Canada the climate is presently too cool to support much growth of *A. flavus*, and aflatoxin is mostly a concern in imported peanut butter or corn products.

The curse of much agriculture in temperate countries is another ascomycete mould, *Fusarium graminearum*, known as red ear rot on maize and scab or head blight on wheat. Although it was rare on my uncle's farm in the 1960s, this fungus is now the main concern in wheat- and maize-growing regions around the globe.

Like ergot and rust, *Fusarium* shoots off its sexual spores onto young plants in the spring, then amplifies in the crop by splashing asexual spores about. On maize, the orange and pink colonies sometimes blanket the whole cob, visible even through the husk, but on wheat its appearance is more discreet. Experienced eyes can pick out the shrunken, tan-colored seeds, which were called tombstone kernels before sensitive public relations specialists renamed them *Fusarium*-damaged kernels (FDK). The slimy orange smears are canoe-shaped spores of the asexual stage. This fungus makes several dangerous mycotoxins, the main one being vomitoxin, or deoxynivalenol (DON). The retching mostly affects pigs, who barf with little provocation, presumably as a defense mechanism. Vomitoxin is now the most carefully controlled mycotoxin in wheat and corn, and a main focus of crop breeders in many countries. Effective regulations mean that DON toxicity rarely affects humans.[22]

A handful of other mycotoxins are monitored carefully at mills, factories, and border crossings. Fumonisins, a family of metabolites produced by *Fusarium verticillioides* and related species, often occur in maize products from warm countries. These compounds were discovered only in 1988. They were overlooked for decades because a peculiarity in the chemical structure let them slip by the methods usually used to detect mycotoxins. Fumonisin molecules look a lot like fatty acids and ingratiate themselves into cell membranes, including those of brain cells. A medley of mysterious human liver and kidney cancers and birth defects are correlated with fumonisin exposure, especially in Africa. A disorder called leukoencephalomalacia, or "mouldy corn disease," causes horses to race frantically around their paddocks, become dopey and torpid, stagger in circles, or lose their ability to walk backwards. Two or three days after symptoms start, they suffer seizures and die. Brain autopsies led vets to diagnose what they called "hole in the head syndrome." After fumonisins were identified as the cause, the disease was prevented by making sure horses weren't offered mouldy food.[23]

Ergot, *A. flavus*, rusts, smuts, and some *Fusarium* species are pathogens that grow inside leaves, stems, or seeds at first, and the plant seems to be healthy for a while. Plant pathologists call this systemic growth, just as doctors use the word "systemic" for hidden diseases in our bodies. The systemic phase is not usually considered symbiosis, but some mycotoxin producers blur the definition. *Fusarium verticillioides*, for example, is sometimes an infective pathogen of maize called ear rot. But it also grows between the cells of stems and leaves without causing symptoms and can then be considered an endophyte. When the affected plant parts are what we want to eat, we consider the result a disease. The plants have a different perspective; the anti-insect metabolites made by an endophyte protect them, at least for a while.

Insects are serious pests on farms. Fungi are often on the plants' side of the dispute, contributing their mycotoxins to the battle. Insects also suffer diseases caused by fungal parasitic symbionts or pathogens. Some of the most curious—and gruesome—fungal relationships with insects involve species of *Cordyceps* and *Ophiocordyceps* called zombie fungi (close relatives of ergot and *Epichloë*). Like a fungal version of the horror film *Alien*, the bodies of infected carpenter ants fill up with hyphae of *Ophiocordyceps unilateralis* that consume them from the inside and then send biochemical signals commanding the insect to climb towards the sun. At the top of a stalk of grass, the fungus punctures the ant's skull, sends up a spectacular colorful stalk, and shoots ascospores into the air in search of the next victim.[24] Could effective biocontrols be developed from such pathogens that would benefit agriculture?

Agricultural Biological Control: *Beauveria* and *Nosema*

The first infective animal disease with a proven microbial cause was a fungal epidemic in silkworm farms in France and Italy. The mummified pupae resembled a popular chewy chocolate candy rolled in sugar, and the name of the candy was adopted as the name for the disease: muscardine. For thirty years in the early nineteenth century, the Italian civil servant Agostino Bassi (1773–1856) moonlighted on failing silkworm farms, hoping to strike it rich by discovering the cause.[25] Through his odd brass microscope, he saw hyphae plugging the lymph ducts of shrunken pupae and realized the white powders on the corpses were masses of spores. His advice to farmers to reduce infections seems strangely familiar to us today. To limit spread, farmers widened the gap between the racks of cocoons and increased the physical distancing between rows of feeding caterpillars. They were told to wash their hands, boil their clothing, and quarantine diseased worms. These recommendations saved the Italian silk industry. The white muscardine fungus was eventually named *Beauveria bassiana* in Bassi's honor. It is the asexual state of one of the zombie fungi.

Most fungal biocontrol products used on a commercial scale are based on moulds that attack pest insects. Originally considered a nasty pathogen, *Beauveria bassiana* emerged as a biocontrol hero in the mid-1930s. It was first considered to be a single species that was pathogenic to a broad range of insects. Now, after taxonomic studies of its DNA, we recognize about thirty species that look very similar but don't interbreed. Some strains are quite specific about what insects they will attack. *Beauveria*'s spores germinate and its hyphae penetrate the insect's exoskeleton, releasing a deadly mycotoxin called beauvericin. Then the mould sops up all of the valuable organic nitrogen bound in the dead host's innards to boost its own growth. *Beauveria* is now used as a biocontrol for many plant pests, like weevils, whiteflies, fruit flies, aphids, and mites. There are hopes that some strains will be useful to control bedbugs.[26]

Swarms of grasshoppers thick enough to block out the sun wreaked havoc in Canada and the United States during the Great Depression, and ravenous plagues of locusts have troubled tropical farmers since biblical times. The billions of locusts in these swarms can cover an area a hundred miles across so densely that they interfere with aircraft navigation. They strip the leaves off crops. Sometimes they leapfrog across oceans, resting on transcontinental barges because they prefer not to fly at night. If there are no boats, they land on the water, drown, and float—gradually creating rafts of corpses on which laggard locusts can land. The green muscardine mould, *Metarhizium acridum*, is now one of the main biocontrols of these pests. Marketed as Green Muscle, the spores are suspended and stabilized in oil, then sprayed from low-flying crop dusters onto the meadows where the larvae live before they molt into the destructive flying form. Both the white and green muscardine moulds grow as saprobes in soil and infect the subterranean locust larvae.[27] And both are also root endophytes of crops like chocolate and coffee, and may share some of the nitrogen they harvest from the insect corpses with their plant hosts. Several species of the microsporidian genus *Nosema*, related to the pébrine disease of silkworms studied by Pasteur, supplement Green Muscle in the biocontrol regimen for locusts. Microsporidia do not grow in agar culture, so spores are grown and harvested from captive colonies of grasshoppers.

Using host-specific plant pathogens as biocontrol agents of weeds is receiving serious attention. Weeds compete with crop plants for nutrients and moisture, but the herbicides used to combat them have similar negative effects as pesticides do. Pathogen species that include strains with narrow host ranges, like *Colletotrichum acutatum*, *C. gloeosporioides*, or *Fusarium oxysporum*, produce spores easily in culture but require special care to formulate into powders or sprays that remain alive long enough to be delivered to a farm for spraying. Some of these products are coming to market.[28]

From Farms to Food

With concerns about factory farming, environmental impacts, and effects on population growth, attitudes towards modern agriculture are polarized. Sustainable (or regenerative) agriculture is the successor to the Green Revolution, with goals of a reduced carbon footprint and less dependence on fertilizers and pesticides. Healthier agriculture is our goal—better for our health, better for environmental health, and less damaging to biodiversity. To feed our growing population healthier food from a smaller land base, we need to apply all we know about fungi, good and bad, that is integral to food production. We all face the same numbers, ten to twelve billion people by the turn of the next century. As advances in medicine allow more people to lead longer lives, these same people also must eat. Agriculture needs to keep pace. And we need to keep looking over our shoulders to see what new threats are emerging. According to the American nonprofit organization known as the Genetic Literacy Project, which has as its motto "Science Not Ideology," nine major plant diseases threaten our food supply today.[29] Of these, seven are fungi or fungus-like organisms, including wheat rust and potato blight.

Although the fungal contribution to agriculture has mostly been harmful, we can adjust farming to reflect the more complete and balanced ecosystem of plants, endo-phytes, mycorrhizae, and rhizosphere microbes that we see in forests. The simplistic, factory-like systems, with their reliance on heavy applications of synthetic chemicals and extensive irrigation, can shift towards managed ecologies based on locally adapted crop plants and phytobiomes, improved water management, reduced tillage, and strategies to reduce diseases and mycotoxins. Many of the current trends towards a return to family farms, organic farming, diversification of crops, locally grown foods, community-supported agriculture (csa), and homegrown versions of once exotic products are moving us in a more sustainable direction. Each step along the path will be smoother if the roles of fungi are considered.

Despite their regrettable talents for interfering with food production, fungi themselves are sometimes our food. There are many more fungal products on our increasingly multicultural grocery store shelves than you might realize.

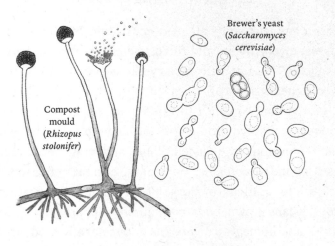

Brewer's yeast
(*Saccharomyces
cerevisiae*)

Compost
mould
(*Rhizopus
stolonifer*)

5 | *FERMENTATION*

Food, Drink, and Compost

MOST HISTORIANS suggest that human towns and cities
originated along with agriculture, which ensured a reliable
source of food. However, counterculture philosophers offer
an alternative: that civilization arose to guarantee a reliable
supply of alcohol. According to this reasoning, our nomadic
ancestors began staying put to grow crops to produce food
for yeasts, including grains such as barley, rice, rye, and
wheat, and fruit such as grapes. Even today, grape cultivation
is far more about wine than it is about food. We are lucky to
find four or five types of edible grapes in our grocery stores,
but every wine shop stocks the fermented juice of hundreds
of varietals on its shelves. By some estimates, the business
around alcohol—wine, beer, sake, distilled spirits—accounts
for 10 percent of the global economy. That's a lot of booze—
and a lot of yeasts. Whether it was beer or bread that led

humans to settle, yeasts really have trained us rather well. You have to question who domesticated whom.

Microbes are expert at transforming unpalatable or indigestible substances into more usable forms. For example, we've seen how various saprobic moulds break down plant matter into humus and soil. In human terms, we think about this kind of process, called *fermentation*, as fungi or other microbes changing raw ingredients into something we like to eat or drink. Fermentation often involves the breakdown of complex proteins or polysaccharides into smaller molecules that are more digestible and offer enticing sensory pleasures. If the microbes alter flavors, textures, or chemistry in ways that we enjoy, compost becomes food. Or drink. Most human societies embrace fermentation in a big way, but the enjoyment of the result is often a matter of taste. One person's delicacy is another person's rot.

In Latin the word for yeast is *fermentum*. "Ferment" also suggests creativity, and it marks a major intersection where humans and fungi collaborate.

From the Beetle to the Bottle: Fermentation by Yeasts

A few thousand species of yeasts occur in nature, but for now we will focus on the famous one. *Saccharomyces cerevisiae* is known as the brewer's yeast or, if you prefer bread to beer, baker's yeast. Both common names refer to the same exuberant species, but the archaeological consensus is that beer came before bread.[1]

In nature, colonies of *S. cerevisiae* live in inconspicuous films on fruit, on bark, and in flower nectar, or they flow in tree sap. Like many other yeasts, this species also streams through the stomach and intestines of insects. Yeasts are well adapted for living, working, and drifting in a liquid world. Although some make hyphae, most remain as pale, single round or oval cells and grow in liquid by budding. This means that the "mother" blebs off a part of herself and gives birth to a clone "child." Both cells rejuvenate and swell up a bit, then might divide again and again every twenty minutes or half an hour. Soon the mother cell is surrounded by offshoots of her former self—grandchildren and great-grandchildren and great-great-grandchildren and so on sharing a cozy, slimy family colony. After three or four days, most yeast cells will die when they run out of food or their swimming pool dries up. The buds that are slurped up by an insect get to carry on. This plant-to-plant, gut-to-gut resort lifestyle was the best of all possible worlds for a yeast of modest ambitions. Then, a new wispy-haired ape showed up, who jumped in grander leaps than any beetle or wasp. From a yeast standpoint, humans were a spectacular vector. So the yeasts climbed aboard and traveled faster and farther with dreams of ever larger vats of sugar water.

Archaeologists detect the chemical fingerprints of sugars characteristic of yeast-fermented beverages on the pottery of many ancient societies. The initial fermentations were probably accidental, but the pleasant tickle on the tongue and yeasty aromas may have inspired our inquisitive nomadic ancestors to settle down to ensure a dependable supply. In medieval Europe, beer was often a part of a worker's wages. It was important to their health. Before the microbial aspects of hygiene were understood, beer was safer to drink than water drawn from streams contaminated with human waste. It was also a dependable source of B vitamins. Brewers cooked grains, especially barley, into a mash, turning the starches into sugars, and then waited. Grapes contain less starch and enough natural sugar that winemakers could simply press the fruit. Both hoped that an infusion of naturally occurring yeasts would settle in and perform their magic.

Today most winemakers and brewers hedge their bets by adding a starter culture of their favorite yeast. Starter cultures are often used in biotechnology—they are kind of like a controlled biological invasion. The starting material is saturated with the desired microbe (or a mixture of several). This saturation overwhelms the resident microbes and blocks weedy moulds from getting a hyphal hold.

Nowadays, commercial breweries are big business. Facilities filling city blocks are optimized to grow as much yeast and make as much alcohol as possible, as quickly as possible. The factories are a cacophony of grinding gears, bubbly golden liquids, and hisses of escaping gas. Copper tanks called fermenters are filled with hundreds of gallons of broth, wort, or mash and connected by a chaotic maze of pipes, switches, gauges, valves, and probes that control temperature, oxygen, acidity, and nutrients for the growing yeasts. Heady aromas saturate the air: the toasted odor of malted barley, the bitter bouquet of hops, the pungency of yeast. In sorting rooms the size of sports arenas, thousands of glass bottles jostle along conveyor belts and are filled, labeled, capped, and then stuffed into plastic crates by robotic hands. Foam gushes out of the bottle necks and bubbles down through grates in the floor.

All this budding and beer making by yeasts makes our own reproductive efforts seem quite modest. I don't like to toss numbers around—biologists are terrible at mathematics—but we should get our heads in sync with the magnitude of the yeast population that shares our civilization. When you

put a yeast cell in a new home, it starts to double: 2, 4, 8, 16...
over and over, and the wonders of exponential growth unfold.
It takes about 5 billion yeast cells to make a bottle of beer, and
humans make more than 50 billion gallons of it every year.
Do the math and that's about 3,000 quintillion (a billion billion, or 10^{18}) yeast cells annually just to make beer—never
mind whatever else they might be getting up to. If each yeast
cell lives for four days, roughly 33 quintillion yeast cells are in
circulation at any one time. That's 80 million times more
yeast cells in our beer fermenters than stars in the Milky
Way galaxy, or more than four billion times more yeast cells
than people on Earth.[2] If you follow the growth of microbes
like yeasts in a test tube, you can watch the sharp rise of the
exponential growth, a leveling off, and then when the available sugar is exhausted, a crash. A brown residue of dead cells
sinks to the bottom. Yeasts need to find a new home almost
every day. They constantly push the limits of growth in a
closed environment. This breakdown of glucose to ethanol
and carbon dioxide by yeasts is the classic example of fermentation in the absence of oxygen, a condition we call anaerobic.

The involvement of yeasts in starter cultures used for beverages and breads was a major step in a domestication process
that's been going on for millennia. Most *Saccharomyces* species, including *S. cerevisiae*, live in nature in China and have
the high genetic diversity there that we expect at the geographic origin of a species. Modern genomic analyses confirm
that unique yeast starters were domesticated to make sake in
Japan, wine in Europe, and a handful of distinct beverages
in other countries. Most yeasts used for modern brewing are
hybrids arising from sexual crosses between *S. cerevisiae* and
other species of *Saccharomyces*, many but not all also native to

Asia.[3] The most widely used beer starter is a hybrid between the Japanese sake and the European wine yeasts.

Although many plants (like maize), animals (like mules), and microbes (like yeasts) readily form hybrids, their off-spring are usually sexually sterile. In domesticated microbes, the guarantee of prolific asexual budding and clonal growth makes hybridization less limiting than it might be in nature. The cells of hybrid yeasts used for beverages often have two or sometimes up to six complete genomes, a condition known as polyploidy.[4] This confuses the cellular machinery needed for sexual reproduction—there are too many chromosomes to sort out. The longer they are domesticated, the more yeasts adapt to the fermentation environment and lose touch with their natural origins. Domesticated brewer's yeast strains are unable to mate or mix further with their wild relatives and seem locked into their genetic tunnels. The artificial selection imposed by their human accomplices ensures that the domes-ticated clone survives and stays more or less the same. With the offer of alcohol in return, this starts to resemble an arm's-length symbiosis.

Most of us have a hard time distinguishing the compo-nents of tastes and smells. If two compounds with unique odors are blended, our noses can be fooled into interpreting it as a third aroma. In fermented beverages, variations in the biochemical fermentation process and secondary metabolism result in the wide range of volatile and soluble compounds that give each product its characteristic odor and flavor. Do you smell green apples? That's probably acetaldehyde, one of the intermediates in the breakdown of glucose to alco-hol (be careful, it's one of the main causes of hangovers). Does your drink smell fruity? That could be an ester, formed by the reaction of alcohol with acids. Medicinal or smoky? Likely a phenolic, either from the wooden cask used to age the liquor or else breakdown products of pigments and other compounds in the original plant. How about rotten eggs? Hydrogen sulfide—proteins contain a certain amount of sulfur that is released when they break down. Sounds delicious, right? Finding the right balance is the art behind the science of fermentation. In nature, the alcohols and volatiles were probably originally an invitation to potential vectors to share sugar. Many creatures attracted to alcohol-infused fruit have the necessary enzymes, called alcohol dehydrogenases, to emerge with their sobriety intact. Animals who don't partake in fermented foods in their day-to-day diet lack such enzymes and react with what we would consider inebriation.

Off-flavors are unfortunately also part of the game. Our noses and tongues are surprisingly sensitive at detecting what we *don't* like. Too much of a good thing, like a metabolite that might be seductive at a low concentration, or the addition of unexpected compounds from a renegade microbe, distorts the end product. Not surprisingly, some people like the result. Particularly in Belgian ales, other yeasts like *Brettanomyces*, as well as lactic acid–producing bacteria (*Lactobacillus* species), are deliberately included in starter cultures to broaden the flavors. But wine lovers are familiar with the cork taint caused by moulds growing in the natural corks used to seal bottles. The guilty metabolite is a chlorine-containing compound called trichloroanisole (TCA), which spoils up to 10 percent of bottled wine. TCA doesn't smell much, but it makes our noses oversensitive to earthy or musty odors.[5] The effect lasts for hours, and sending the bottle back is the only option. Reducing cork taint is the main reason screw caps are now commonly used on wine bottles.

Beer is the primary product of any brewery, but sediments from the fermentation tanks have a second use. Dead yeast cells, rich in B vitamins and glutamates, are filtered out and compressed into cakes for feeding livestock or transformed into food for humans. The love-it-or-hate-it sandwich spread Marmite (called "Vegemite" Down Under, and "Cenovis" in the Alps) is made from spent yeast that is heated, cooled, salted, and concentrated into a thick paste. To make single-cell protein (SCP), a fermentation with brewer's yeast is altered to favor protein production and cell mass instead of ethanol. "Torula" or "nutritional yeast" is a similar product based on a different fungus, *Cyberlindnera jadinii* (often called *Candida utilis*). The cells are harvested at peak health and dried into a powder used as a nutritional supplement for humans and animals. Although some people sprinkle these powders onto cereals or into yogurt, not everyone enjoys the bitter, yeasty taste.

Fermented beverages can themselves get contaminated and undergo a second fermentation. For example, ten to twenty species of yeasts and bacteria turn the alcohol in wine or cider into acetic acid. If the wine is contaminated accidentally, the result is often sour or musty. Starter cultures for vinegar assure a happier outcome. The resulting wine vinegars are used as preservatives and valued flavor enhancers. Similarly, kombucha is brewed tea that is fermented by a starter culture of yeasts (often brewer's yeast) and lactic acid bacteria called a "mushroom" or SCOBY (symbiotic culture of bacteria and yeast). The result is a fruity drink that is low in alcohol (0.5 percent or less). Having watched tiny protists zip around in homemade kombucha starters through a microscope, I'm usually not eager to drink the result. But quality control of commercial kombucha should allay such concerns.

A Slice of Life: Yeast-Leavened Bread

Although brewer's yeast is lauded for its talent with liquids, it is equally celebrated for its abilities with dough. The first known loaf of yeast-leavened bread came out of the oven only about four thousand years ago, long after flagons of beer were commonplace. Unleavened bread, made without yeast, is flatter and less spongy, and dates back an additional millennium. Perhaps foam overflowed from a beer vat into a batch of bread dough left lying around by a distracted baker, who was pleasantly surprised by the light airiness and homey aromas when the loaf was baked.[6]

Bread is sort of beer turned inside out. The fermenters used for beer and wine are closed systems with inlets and outlets that in some ways mimic human stomachs. Bread dough is a more open system. Instead of precooking ingredients and confining them to a tank, bakers spread raw ingredients on the counter or dump them into a bowl, add a yeast starter dissolved in liquid with a little sugar, and then mix it all up on the counter. The excited yeast, unaware that it is on a suicide mission, foams the batter up with carbon dioxide and buds madly until the mixture is ready to be kneaded into loaves. Gluten—the stretchy, sticky proteins in milled grain—captures gas bubbles in pockets of the rising dough. During baking, they expand and give bread its spongy texture.

Commercial bakeries everywhere use only a few yeast starters. They are the same species as brewer's yeast but different strains, and sometimes hybrids. This explains why most industrial or commercially baked breads taste so similar. You can buy the same clone in your grocery store, compressed into a dense paste or dried into shiny granules. Most kneading in bakeries is done with dough hooks inside sterile mixing bowls. Unless other ingredients are added, the main microbial activity is only the original *Saccharomyces*. The fermentation lasts just a few hours, so microbes in the air or on the counter have little time to fall into the dough and alter its properties.[7]

Hand-kneaded breads are more diverse, and artisanal breads really do taste different. The community of microbes in a sourdough starter is much more complex than the monoculture used for ordinary bread. Sourdough starters might begin as a clean blob of batter, but wild and domesticated yeasts and bacteria from bakers' hands and the surrounding air are kneaded into the mix. Lactic acid–producing bacteria give the "sour" to the dough. Over time, the starter culture settles into a stable ecosystem, and the hands of the baker and the cracks and corners of work surfaces tend to host the same happy blend of microbes. Personal starters are precious and often have pet names, like Jane Dough or Rye Breadbury. Bakers keep their starters in the fridge or on the windowsill, and transfer a bit into a new batch of flour and water every few days to keep it invigorated. Sometimes a cranky yeast or bacterium invades the starter and reduces it to a weary, stinky paste. Then, a contingency plan, like sharing the starter with a trusted family member, allows the valued culture to be rescued. One enthusiastic Belgian baker curates a starter dough library, preserving the diversity of sourdough communities for research and posterity.[8]

Say Cheese: Mould Fermentations of Dairy Products

Bread and alcohol fermentations are variations on the *Saccharomyces* yeast theme, but many fermented foods involve moulds. With cheese, the moulds either grow inside the curds or form the rind on the outside. The flavor, taste, and texture depend on what kind of milk you start with (cow, sheep, goat, or something more exotic), how you mix and pack the curds, what fungi or bacteria you add or let settle, and how long you are prepared to wait. Even the breed of the cows and location of the farms supplying their food affect the result.

Rock paintings suggest our ancestors stumbled across cheese making about six thousand years ago. Caves still feature in cheese folklore—the French and Italian versions of mould-enhanced blue cheese, for example, share similar origin stories. A naive young shepherd or apprentice is distracted by an attractive individual of a compatible mating type. During their dalliance, their lunches languish in a cave. Eventually they return to find their sandwiches gone mouldy. Having gone without food and who knows what else for several days, the pair eat the fuzzy cheese and declare it delicious (the French Roquefort) or blithely use it as a starter for a new cheese in the hope that no one will notice the funny color and flavor (the Italian Gorgonzola). The modern warehouses of high-volume producers duplicate the cool, humid conditions of caves, with floor-to-ceiling shelves holding aging wheels of cheese. But artisanal makers often still mature their wares in natural caves or abandoned wine cellars.

Modern industrial cheese facilities are similar to large breweries. Plump tanks of coagulating curds sit among assembly lines for compressing, shaping, and packaging the product. Curds comprise cloudy wisps of protein and lipid that appear after an enzyme called rennet is stirred into milk or cream. Most rennet comes from the stomach lining of slaughtered cattle. Vegetarians, therefore, prefer cheeses initiated with plant enzymes or fungal rennet extracted from cultures of the zygomycete *Rhizomucor miehei*. After curds are squeezed into balls or sliced into slabs, the fermentation and succession proceed. The interiors of most cheeses are fermented by lactic acid bacteria, which occur naturally in unpasteurized milk. In some countries, the risk that the potentially deadly bacterium *Listeria* might appear during aging requires that the milk used for cheese be pasteurized, which means the liquid is heated to between 130 and 145 degrees Fahrenheit. This process wipes out the original microbial population in the milk, and the cheeses must be sown with a starter culture.

Hard cheeses slathered with a layer of molten wax or wrapped in plastic tend to remain mostly bacterial, and any mould growth is considered contamination. But cheeses with natural rinds are covered with a thin, tangled skin of the interwoven mycelium of several moulds. These are still often aged in caves, and spores of naturally occurring moulds are stirred up from dusty floors or flutter down from spiderwebs to settle on the slabs. DNA barcoding surveys reveal hundreds of fungal and bacterial species in the curds and on the rinds of these styles of cheese. Each has a role in creating the individual flavor, smell, and texture of the product—sometimes working in tandem, sometimes in succession—and the exact mixtures are specific to individual caves or farms. It was only recently discovered that the mould *Geotrichum candidum* provides most of the white mycelium in these rinds. Its hyphae meander over the outside and it has a yeast-like form that sometimes penetrates towards the middle, breaking down some of the lipids and proteins into tasty or pleasant-smelling smaller molecules. For traditionally aged cheeses, individual caves often have resident *Geotrichum* strains that contribute a unique flavor. If a different strain invades, the expected flavor might be spoiled.[9]

A less welcome fungal visitor on some rinds is *Mucor mucedo*. It makes a gray to green fuzz known as "hair of the cat" (*poil de chat*). Its funky flavors are acceptable on some kinds of cheese, but for varieties with more nuanced aromas the entire brick must be tossed out. And if this contamination occurs too frequently, the cave is abandoned. The "flower of the moulds," *Trichothecium roseum*, turns some cheeses pink, but otherwise its role (and whether it might make mycotoxins in cheese, as it does in agar culture) is poorly understood.

More than 1,800 cheese varieties grace the world's kitchens. The mould-based types in most Western refrigerators are blue (or *bleu*) cheese and white Camembert or Brie. The star fungi in both are species of *Penicillium*, which make high levels of enzymes that break down fats and proteins. Of these two types, blue cheese is much older. Its varieties all involve *Penicillium roqueforti*. Cheese makers aerate the aging cheese by stabbing it with pins or spikes, and *P. roqueforti* fills those holes with millions of spores that form dramatic dark blue or green veins. Gorgonzola is the oldest kind, made for more than a thousand years in towns around Milan, Italy. The English made a similar cheese in the town of Stilton starting in the 1820s.

Blue cheese is not to everyone's taste. The flavor is salty and sharp, and the smell comes from organic acids. Perhaps these volatile compounds repel mites, which are seldom seen on the product although they often occur on other mould cheeses. The smell certainly repels some humans. If you have never tried blue cheese, just consider rancid butter (butyric acid) stirred up with old bananas (ketones) and spread onto a goat (hexanoic acid), and you can imagine the odor. I love blue cheese, especially with walnuts and pears and accompanied by port. Lots of people agree: world exports exceed half a billion dollars per year.[10]

Brie and Camembert are named for the cave-riddled regions of France where they are produced. During the French Revolution, according to legend, Marie Harel (1761–1844) sheltered a renegade priest from irate mobs. During the cleric's exile, the pair passed their happy hours with clandestine cheese making. Their major innovation was to omit cream from the traditional recipe for Brie, and they adopted the name "Camembert" for the result. As a marketing ploy, they crafted those little wood-wafer boxes to transport their mini-rounds for sale. The fermenting agent responsible for the white rind and smooth white interior of both was always assumed to be a relatively pure culture of the white-spored *Penicillium camemberti*. Now it is clear that the white-spored *Penicillium caseifulvum* and *Geotrichum candidum* are also often involved. The combined fermentation talents of the mixture of moulds and lactic acid bacteria result in a subtle, slightly nutty odor—but not everyone enjoys the accompanying ammonia note.

These two white *Penicillium* species seem to be domesticated and are rarely, if ever, found outside of cheese-producing areas. The green *Penicillium commune*, which spoils cheese and yogurt, is probably their wild ancestor. However, the blue cheese species, *P. roqueforti*, maintains a diverse population in the wild, although specific strains were domesticated as starter cultures. In the wild, this species is quite rambunctious, and colonies (which can be the size of soccer balls) sometimes swell up in grain silos and poison any cattle that accidentally eat them. Ominously, two of the main cheese species sometimes produce mycotoxins. When it grows in silage, *P. roqueforti* produces PR toxin as well as mycophenolic acid and a few other questionable compounds.

Under some conditions *P. camemberti* makes cyclopiazonic acid. All can poison any cattle that accidentally consume them. Why do they poison cattle and not humans? When the moulds grow on dairy products, either the toxins are not produced or the ammonia emanating from the cheese breaks them down.[11]

My mother felt that these fancy French cheeses were unnecessary; there was only ever cheddar in her refrigerator. She maintained a strictly meat-and-potatoes kitchen, typically augmented by frozen or tinned vegetables. The only foreign food we ever ate was the occasional takeout from one of the Americanized Cantonese restaurants in town. The little plastic packets of soy sauce that came with the steamed rice were usually just stuffed into the corners of the fridge or wedged in beside the dusty bottle of the same at the back of the spice cupboard. It wasn't until I went to university that I discovered soy sauce is a fungal fermentation product and a gateway into the world of delicious Asian fungal foods.

The Feast in the East: Soy Sauce and Other Adventures of the Kōji Mould

Soy sauce was invented in China about two thousand years ago as a cheap substitute for expensive rock salt. Buddhist monks traversing the Silk Road carried their recipe across the continent, and chefs modified the original to suit their own tastes, resulting in local variants like tamari, ketjap manis, and light and dark soy sauces. Although artisanal soy sauce is still produced as a disciplined, monkish activity, industrial factories make most of the 2.5 billion gallons sold every year.

Unlike most liquors, breads, and cheeses, soy sauce is made using a two-stage fermentation process.[12] For the first step of many fermented Asian foods, various grains undergo an aerobic fermentation by the mould *Aspergillus oryzae* to produce an intermediate called kōji. The mycelium overgrows and penetrates a mixture of cooked soybeans or grains for a few days at 85 to 100 degrees Fahrenheit. The mould's enzymes break the starches into simple sugars and the proteins into amino acids, especially glutamic acid. Natural glutamic acid contributes rich flavors to most fungal foods, from soy sauce to Marmite and mushrooms. It is also the chemical backbone of monosodium glutamate (MSG), the unfairly maligned umami-enhancing condiment that sits beside salt and pepper in many Asian restaurants.

It is hard to find a more remarkable example of domestication than the kōji mould. The aflatoxin-producing fungus, *A. flavus*, is considered the wild ancestor of the domesticated kōji mould, and research shows that their genomes are 99.5 percent identical. They can hardly be distinguished from each other, even with the best microscopes, yet one is useful to humans and the other can be toxic.[13] Domestication happened many times, over and over again in different regions. The domestication of microbes seems haphazard, though, when compared with the more familiar examples of animal domestication. Originally no one realized that living things were involved in processes like cheese making or kōji, nor did they understand that their selections altered or removed specific genes. But kōji domestications stabilized a set of about 150 genes involved in breaking down starch—a genetic modification that correlates with what the innovators were trying to encourage. The most surprising genetic change is that in many kōji strains, the aflatoxin genes have disappeared completely. They do remain in some strains but are silent and the toxin is never made.[14] Most kōji strains make far fewer spores than wild forms of *A. flavus*. Perhaps such strains were selected because the result appeared less mouldy. Or perhaps the fungi don't need as many spores to disperse themselves because humans do that for them.

Each geographical region has a characteristic kōji, based on indigenous strains of *Aspergillus* and other benign moulds and bacteria, which provide nuance to the eventual products. Traditional Japanese-style soy sauce, for example, starts with a blend of steamed soybeans and baked wheat naturally inoculated with kōji spores from the air or added as a starter. Tamari is a variant that traditionally omits wheat.[15]

In the second stage of fermentation, the kōji-fermented slurry from the primary fermentation is mixed into water containing up to 20 percent salt. This goop undergoes a second, mostly anaerobic fermentation. Factories have massive stainless-steel tanks and age the liquid for four months. Monasteries use wooden tanks and leave them for two to four years. The high salt concentration kills most microbes by drawing much of the water out of their cells. Only microorganisms that thrive in low-moisture environments—xerophiles—will grow. Among these is a bacterium with the memorable name *Tetragenococcus halophila*, which transforms some of the sugars into lactic acid and lowers the pH to about 5, roughly the same acidity as coffee. Then *Zygosaccharomyces rouxii*—let's call it the salty yeast—starts to grow. It makes low levels of alcohol, less than 2 percent, and a blend of metabolites—mostly glutamates and breakdown products from other amino acids—that contribute malty, maple syrup, caramel, cooked potato, curry, and various fruity notes to the bouquet. When the fermentation and aging ends, the sauce is decanted and pressed out of the sludge. Then it is pasteurized and bottled. The felt-like cake of microbial cells and plant fibers left behind is dried and fed to cattle.

The kōji mould is used as a starter for various primary fermentations that break down starches in rice, wheat, soybeans, and even sweet potatoes. Then they are further fermented to make mirin, miso, pickles, sake and other liquors, and many other traditional Asian foods. Kōji has spilled over into the Western world in surprising ways. New York chefs now use kōji mould to artificially dry-age expensive beef steaks. After only two to three days, they achieve aging effects that would normally take forty-five days of hanging. You can find instructions on the Web to try this at home, but it is not for the faint of heart.

Dessert With a Beverage: Chocolate, Tea, and Coffee

Because I want to leave you with good vibrations about fermented fungal foods, let's complete what until this moment you may not have realized is a sacred fungal trilogy: wine, cheese, and chocolate. Chocolate making is an old biotechnology that dates back to about 450 BCE. The Aztecs considered cocoa a gift from the god of wisdom, Quetzalcoatl, who was expelled from paradise for sharing such a heavenly delight with lowly mortals. At that time, most chocolate was mixed with spicy or bitter ingredients and used in drinks. Not until 1847 did the confectioner Joseph Storrs Fry II (1826–1913) invent the chocolate bar in Bristol, England. Chocolate truffles have blessed us for less than one hundred years... and were named for their resemblance to fungal truffles.

Chocolate starts off as orange or brown pods that dangle off the branches of cacao trees like deflating balloons. A slice with a machete and the "beans" and white glutinous slime inside the pods can be scooped into large bins, where they are covered with banana leaves and left to ferment naturally. Brewer's yeast is one of the first fungi to colonize the mess.

It produces a bit of ethanol in the mildly anaerobic pulp, and then releases enzymes that break down the pectin that glues plant cells together. As the fermentation proceeds, the temperature rises above 100 degrees Fahrenheit. Then a succession of other yeasts (varying from country to country, but usually species of *Candida*, *Kloeckera*, and *Kluyveromyces*), lactic acid bacteria, and acetic acid–degrading bacteria blossom in the pulp. This succession of yeasts and bacteria stimulates chemical reactions that transform the bitter raw cocoa beans and release up to four hundred flavor compounds. These volatile fruity and floral esters and alcohol metabolites give chocolate its distinctive flavor and aroma. Over the past decade, starter cultures were developed to offer more control over the fermentation and are used by some small producers to ensure a more consistent product.

After about a week, the fermented beans are washed, dried, and roasted at 250 degrees Fahrenheit to kill off the remaining microbes, caramelize the residual sugars, and concentrate the flavors. The outer shell of the bean is removed and the soft tissue inside is ground into paste, pressed, and mixed with other ingredients. Then the texture and flavor are tweaked before the mixture is molded into shape. Eventually, what you think of as chocolate finds its way to you, to melt on your fingers and slide down your throat with a cup of tea or coffee.[16]

Although regular black or green tea does not involve microbial fermentation, traditional Chinese Pu-erh tea, naturally fermented in an aerobic process by *Aspergillus niger* and a yeast-like fungus called *Blastobotrys adeninivorans*, will satisfy your fungal muse—assuming you can afford the luxury price.[17] You can try to pair its floral, smoky, sweet, or sour notes with your favorite chocolate. Or you can indulge in another beloved fungus-modified beverage, coffee, instead.

Coffee fermentation has a lot in common with chocolate. Coffee "beans" are twinned inside the red "cherries" that cluster along branches of coffee trees. Before roasting, farmers rub the skins off the beans, wash them, and let them sit for several days. The anaerobic fermentation is dominated by a yeast called *Pichia nakasei* and some lactic acid bacteria, with *Candida parapsilosis* sometimes showing up later. Lactic acid and acetaldehyde are the main yeast metabolites that build up, but volatile organic acids, alcohols, and esters also concentrate in the dried beans. When they are roasted, these metabolites join the caramelizing sugars and blossom into the seductive flavor and aroma of the brew. They also make the product acidic and less likely to spoil.[18]

You should enjoy your coffee while you can. Coffee crops are under serious threat—from climate change, low genetic diversity, and the invasive coffee rust *Hemileia vastatrix*. This rust followed coffee trees around the globe as imperial powers fell in love with the brew and planted it in far-flung countries. The disease was first seen in Sri Lanka in the mid-nineteenth century, where it destroyed the coffee crop. It was then that the English lost their coffee supply and adopted tea as their afternoon drink. Now it has finally caught up with the prized crops in South and Central America, with yield losses of 90 to 100 percent. Coffee breeders hope to save us by generating resistant cultivars.[19]

To Eat or Not to Eat: Rotting Food or Edible Compost?

From a fungal point of view, there's no difference between human food and human garbage. Fungi don't discriminate between what you store in your refrigerator or pantry and the scraps you denounce as compost; they just help themselves to nutrients from whatever organic matter they find.[20] The green mould colonies powdering drying slabs of hard cheese or old bread are warning signals. You can't just cut the visible mould off and carry on with your snack. The haze of spores doesn't tell you how far the hyphae have penetrated into parts that look clean—there could be mycotoxins in the soft, faded zone where exuded fungal enzymes are breaking down fats and proteins. So the rule of thumb is "Don't eat mouldy food." And don't feed it to the dog—and if you are a farmer, don't feed it to your cattle either. Every year, animals die from eating food contaminated with mycotoxins made by some of the same *Penicillium* species that spoil human foods.

According to the Food and Agriculture Organization of the United Nations, one-third of food spoils before it is eaten, an amount large enough to feed 600 million people.[21] Most of this loss is caused by fungal biodegradation or contamination with mycotoxins. Some of this loss occurs while crops and livestock are being processed or stored on farms, but most of it happens at home. The percentage has been constant for all our lives.

To prevent moulds from spoiling food before we eat it, we use the same physical and chemical logic as we use to protect harvested grains on a farm, or trees in a forest. To grow, living things need tolerable temperatures and acidity, and sufficient oxygen, nutrients, and water. Refrigerating food below 50 degrees Fahrenheit slows spoilage, but some moulds still grow at low temperatures. Fresh food in your refrigerator is an invitation for hungry moulds that tolerate cold temperatures. Every time you open and shut the door, their spores are pulled inside by air currents. They eventually sprout and grow onto crumbs hiding between the cracks or spread onto other unwrapped food. A pinhole in the lid is enough to let one spore of the fat- and protein-loving *Penicillium commune* squeeze into your yogurt container. Then you find a powdery green colony floating inside when you peel the foil back. Oranges covered by sneeze-inducing green rot (*Penicillium digitatum*) and bruised apples with mycotoxin-producing blue mould (*P. expansum*) often fall to the bottom of the fruit drawer. This spoilage is predictable and explains why most foods have a "best by" date.

Freezing food stops all mould growth, though it's not always the best choice for some foods. Canning and vacuum packing eliminate oxygen, as does backfilling plastic packages of salads, potato chips, and fresh pasta with nitrogen gas. Still, some fungi get by with just a bit of oxygen. Drying foods removes the water that fungi need to grow. Cells immersed in salty or sugary liquids expel water, trying to balance inner and outer concentrations. But because they can make glycerol in their cells and tighten pores in their cell membrane to reduce water leakage, moulds known as xerophiles can often cope with the "virtual dryness" of salted or sweetened foods. Xerophiles are among the most common contaminants of preserved food and stored grains. In almost any pantry, xerophilic *Aspergillus glaucus* and related species run rampant in jams, sweet pastries, or fruit juices (even sometimes in sealed containers, because of pinpricks in the packaging or inadequate cleaning or pasteurization of the fruit). Through the microscope they look like yellow, egg-shaped geodesic domes filled with spores shaped like flying saucers.[22]

If we want food to be toxic to fungi but not to us, we use food preservatives. Traditional methods like salting and pickling food in acids like vinegar are not always effective, however. Salting makes foods attractive to xerophiles, and most fungi like acidic conditions too. In fact, they routinely release organic acids themselves as a way to slow down competing bacteria. Hundreds of preservatives are in use, some natural and some artificial, and they are usually much safer for consumption than the chemically complex synthetic molecules used on farms as pesticides. Preservatives like benzoic acid and sorbic acid interfere with the fungal metabolism of glucose. Even if spores land and germinate, the new hyphae can't use the sugar and they starve. Both are flavorless "natural" metabolites discovered in the 1500s after the alchemist Nostradamus noticed (in his pre-prophecy days) that cloudberries and mountain ash fruit last a long time before they start to rot.[23] And you hardly see *Neurospora crassa* anymore because of the calcium propionate used in commercial baked products—the species is now so rare in bakeries that its former common name "red bread mould" is hardly ever used. But it is still comfortably at home in nature, forming powdery orange blooms on charred tree trunks after forest fires.

Using several strategies at once to preserve food works best. Each approach slows a fraction of the fungi knocking on the door. Putting two or more together cuts down the list of potential contaminants still further at each step. But a few moulds always find their way through. The cold-tolerant, acid-loving, sorbic acid–eating, moderately xerophilic, and oxygen-indifferent blue cheese mould *Penicillium roqueforti* is one of these multitalented moulds that is quite difficult to discourage. It's a good idea to keep that blue cheese well wrapped up in the fridge.

Oddly, another solution to preserving food is just to let nature take its course and evaluate the consequences. It may not be comforting to think of the world as one big compost heap, but as we discovered over the centuries, some end products of biodegradation processes like fermentation are quite tasty. And others are not, and can be hazardous. In any compost, a succession of moulds take their turn as organic matter transforms and breaks down.[24] When food is moved out of the refrigerator and left at room temperature, spores that carry over from their time on the farm, or settle from disturbed house dust, germinate and start to grow. After a day or two, the fruits and tomatoes collapse from hyperactive digestion by *Rhizopus stolonifer*. It likes anything with easy sugars and abundant water. *Rhizopus* grows so rapidly that it often spills over the edge of the compost bucket like a mist of wet cotton sprinkled with pepper. When you watch it with a microscope, the hyphae jump around like the runners that stretch out across the soil from one strawberry plant to another. Jutting out from the hyphae are clusters of what look like oversized pushpins. Black specks that flake off their helmet-like headgear resemble charred skin, freeing clouds of wrinkled gray spores to waft into the air. When the compost starts to dry out, colonies of *Penicillium* and *Aspergillus* cover everything with green, yellow, and black powder. Banana peels and chunks of cauliflower and potato are full of cellulose, starches, sugars, and oils that fungi love. Which moulds appear on your compost will vary according to what foods you started with, where you live, and the time of year.

Dangerous pathogens are unlikely in household compost, but people with allergies or asthma will suffer as the millions of spores float into the air. In some composts, surface microbes respire aerobically, but those buried in the center may be somewhat anaerobic. Mixing up the compost favors the aerobic microbes and reduces odors. In large-scale composting facilities, wood chips piled outside pulp and paper mills, or manure and straw heaps on farms, the energy released by microbial metabolism heats the inside of the pile to human body temperature or beyond. Then heat-loving moulds called thermophiles appear in the succession. If you inhale their spores, there is a risk they may grow inside your lungs. So especially in the summer, get the compost out of the house at the end of each day.

If you plan to use your compost to produce more food, you want it to be crumbly and fiber- and nutrient-rich, very similar to the humus produced on a forest floor. Leaving the results to the randomness of the natural succession is one option. Another is to add a starter culture that will help break down undigested plant matter. Mature compost still often contains a lot of cellulose, but we can use it to grow edible fungi that degrade cellulose and further reduce waste.

Compost is used to farm the familiar edible mushroom *Agaricus bisporus*—known variously as button, cremini, coffee, or portobello, though it is all one species. In commercial operations, instead of the traditional composted horse manure, a blend of straw (mostly cellulose), gypsum, and nitrogen-rich farm waste (like chicken droppings) is left to ferment outside in long steaming berms for one or two weeks. Then massive backhoes shovel it onto the shelves inside Quonset huts. When the compost has cooled, a starter culture, spawn of pure *Agaricus* mycelium growing on cooked grain, is stirred in.

The most common cultivated oyster mushroom, *Pleurotus ostreatus*, is also propagated using waste. Plastic bags are filled with sawdust, straw or other agricultural waste, or even newspapers or coffee grounds, then inoculated with a starter and stacked on shelves in temperature-controlled barns. After the white mycelium binds the substrate into a nougat-like mass, holes are poked in the bags and the mushrooms burst forth in clusters after a few days. Kits to grow them at home are quite popular.[25]

As unappealing as it might sound, compost itself sometimes becomes food. In tropical Asia, soybeans thrown away after being boiled to make tofu are often quickly overrun with *Rhizopus oligosporus*, a thermophilic relative of the black compost mould. The result is tempeh, a protein-rich cake of fermented soybeans, originally sold in Indonesian farmers markets and until recent decades hardly known in the West.[26] It was probably a chance discovery of a peckish Indonesian farmer nibbling on the fuzzy edges of discarded mash woven together by mycelium. In the natural succession, the spores of *Rhizopus* settle from the air onto cooked bean patties, which are then wrapped in hibiscus leaves. Hyphae race through the packed soy, altering the proteins to make them more digestible. Nowadays, commercial tempeh producers in North America and Europe compress boiled beans into plastic moulds and add *Rhizopus* spores directly. This addition short-circuits the somewhat random natural colonization and succession and improves the chances that an edible product will result after a few days.

As part of my introductory mycology course at university, we made tempeh from scratch. Our efforts were directed by a grad student with a roguish twinkle in his eye and a carefully groomed mustache that gave him a certain chef-like panache. Each of us in the lab group was assigned to monitor the boiling of a beaker full of soybeans that was held on a retort ring above the blue flame of a Bunsen burner. When we were sure the beans were soft enough, we pressed them into half-inch-thick cakes in Petri dishes and waited for them to cool. Then we sprinkled them with a homemade starter of rice flour mixed with spores of the tempeh mould, stacked them in a growth chamber set to tropical temperatures, and left them to ferment. Several days later, the grad student invited us back to the lab. He slit the mouldy soybean cakes with a scalpel freshly removed from a sterile foil packet, then sautéed the tempeh slices in a frying pan on a hot plate that was usually used to melt agar media. We all gathered around to sample the result. Using that standard lab accessory, a toothpick, I speared a little piece. The tempeh was dense and crumbly, with an earthy, almost meaty basal note. It was inoffensive, but I wondered why anyone would eat it. Still, at a time when cooking and eating in a science lab were forbidden—and the whole notion of consuming the results of research experiments was frowned upon—I was charmed by this illicit feast. It was a far cry from the stodgy menus of my childhood.

The lesson that waste becomes food and composting becomes fermentation took root. I began to understand that whether something is compost or food is a matter of perspective. What is food for a fungus can also be food for us. What is food for one creature might be garbage to another. But the carbon keeps cycling along, which is essential for the sustainability of our planet. Compost is a thriving microbial ecosystem, and we should keep our eyes open for more happy accidents that might transform waste into new tasty foods or rich, productive soil.

Whether you are reading this book with a selection of fungal delicacies in hand or not, chances are you are seated in a comfy chair in your home or office. These indoor spaces protect us and provide secure caches for our food. Fungi like these warm and sheltered environments too. Every time we leave our windows open, let our pets in or out, take off our muddy boots after a nature walk, or haul produce in from our gardens or farmers markets, moulds, yeasts, and soil fungi follow us inside. Our buildings are fungal fermenters of a different kind.

Stachybotrys
chartarum

Indoor
spores

6 | *THE*
SECRET HOUSE
Fungi and the Built Environment

WHEN I STEP into a house for the first time, I scan the surroundings for interesting fungi the way someone else might search out an interesting conversationalist.[1] This quirk is more socially acceptable outdoors. People get uneasy when I try it in their homes because they'd rather not know there are moulds in plain sight. The particles bobbing in sunbeams. The haze of dust on countertops and the floor—probably spores. The faint water stains on the ceiling tiles. The pink smudges in the grout around the sink in the bathroom. The dark splotches on the door gasket of the fridge. The wilted houseplants covered with brown specks. These all signal growing moulds. And the musty odor seeping up from the basement? Microbial volatile organic compounds (MVOCs). These are mixtures of aromatic mould metabolites like 2-octen-1-ol (said to have a green, fatty smell), and geosmin (earthy or musty), which is mostly produced by fungus-like bacteria called actinomy-cetes but also by a few common moulds.[2]

Building artificial shelters separated humans from their surroundings and provided security against extreme temperatures, strong winds, and storms. A modern building is kind of an inflated version of a body. Its wooden skeleton and the air, heating, and water circulating systems are hidden away, while the walls are the skin protecting our bodies and possessions. During the energy crisis of the 1970s, when the price of oil skyrocketed, the way we made buildings for temperate climates changed. To reduce heating costs, we increased insulation and reduced ventilation. The result was buildings with increased humidity and warmer air, both developments that made fungi feel more welcome.

Like farms, modern buildings are artificial ecosystems, but they are even more divorced from the usual patterns of nature. We didn't plan for all of the creatures and microbes to move in with us, mostly without us noticing. Often the moulds, insects, and rodents that thrive in our warm and dry bedchambers, living rooms, and offices are native to deserts. More tropical or humidity-loving life-forms congregate in our bathrooms, kitchens, and laundry rooms, where dripping taps and billowing steam create a rainforest-like environment. Some of the fungi in our buildings blow through the windows or are carried in on food. We don't need to worry much about most of them: small colonies that grow indoors are normal and play a part in training our immune systems to tolerate the impurities of our chosen habitat. It's profuse growth that sets off the alarm bells. Indoor mould growth that increases the number of spores in the air is known as *amplification*. Modest amplification is tolerable, but we don't want concentrations of airborne spores or MVOCs to get too high.

The meaningful biological divisions among fungi in the built environment are between moulds causing decay; moulds that amplify in wet places; moulds that amplify in dry places; and moulds causing allergies. Each zone of a house has its own guild of fungi that affect us in various ways.

Under the Floor and Between the Walls: Wood Decay and Sap Stain

In most homes, you don't see the wooden frames that support the roof and floor and separate the rooms. But if you walk into a basement or garage, often you will see the vertical wall studs and horizontal floor joists. The first house I bought had an unfinished basement, and I was worried about the condition of the beams. No suspicious polypores or mushrooms were evident, but on some of the joists near the rock foundation, I could scrape away a quarter inch of softened fiber with my fingernail. That gentle corrosion is soft rot, a superficial decay often associated with the asco *Chaetomium globosum*. Its dark, burr-like sexual structures are covered with hooked hairs designed to catch on insect legs or rodent fur. Severely rotted wood is soft or crumbly, but sometimes lumber is badly weakened before it is noticeably rotten. When I stabbed a pocket knife into the solid wood, a long sliver lifted away from the surface. That was good news; the cellulose below the soft rot was still intact. If the cellulose fibers had been snipped by a fungal cellulase, the fibers would have cracked in the middle, indicating that the wood was already weakened. The building inspector confirmed that the strength remaining in the logs was sufficient to support a house of twice the size.

The wall studs in the garage were exposed and had patches that looked as if they were shaded with a pencil or dabbed with india ink. The smudges are a consequence of grayish or pale brown hyphae of saprobic *Ophiostoma* species or other dark moulds penetrating the wood fiber. The hyphae tunnel through pores and spread along the water-transporting vessel cells of incompletely dried sapwood, sometimes softening the cellulose. On dry lumber, this sap stain (or blue stain) is superficial and easily planed off; it has little effect on wood strength, but many people consider it unsightly. The common sap-stain species on lumber are not pathogens.[3] However, some serious tree diseases also produce a deeper, darker version of blue stain. The destruction of conifer forests in the Pacific Northwest of North America by the mountain pine beetle left lumber companies searching for a use for millions of pine boards with deep blue stain. This richly patterned wood was rebranded as "blue denim pine." It is now used as a decorative accent on exposed beams and panels in many public buildings in British Columbia and the rest of Canada.

Most structural wood is kiln-dried to reduce water content below 20 percent, and if it is intended for furniture, well below 10 percent. By the time lumber is assembled into houses, the wild wood-rotting fungi should be dead. In soft rot and blue stain, the core of the wood remains sound. Both defects are often a sign that the lumber was once wet, or may still be wet and then vulnerable to ongoing decay. Decay is a real enemy in buildings, and it usually happens when wood is rewetted. If plumbing leaks or condensation drips from cold water pipes, gallons of water can drip onto floor joists. Vapor barriers trap internal moisture and stop the lumber from drying out. The wood fibers swell up and spores that land on the wood during construction germinate, or mycelium edges through cracks in the foundation from the soil outside. The invisible hyphae of wood-rotting polypores and mushrooms get to work, and decay starts. Most leaks are small and difficult to locate.

Building inspectors use moisture meters to track down leaks behind wallboard.

Serpula lacrymans is a common wood-decay basidio that causes dry rot. In nature, this fungus is only found high in the mountains of Asia, but it hitchhiked to the New World hundreds of years ago on the same wooden ships that ferried humans. It needs just enough water for its spores to germinate, and it harvests this water by breaking down cellulose and transferring it back through the mycelial cords to the dry parts of its colonies. The expanding brown fans of droplet-oozing mycelium and winding rhizomorph-like mycelial cords give *Serpula lacrymans* its name, which loosely translates as "the serpent with tears." The rippled rusty-brown crusts spread along basement floors and ceiling joists at about 3 inches per day, often hidden from view, and start throwing out clouds of spores. These blow through heating ducts and settle in a rusty powder on the floor—the only clue homeowners may get before their floor, or sometimes the whole building, collapses.[4]

Finding environmentally friendly and human-healthy ways to preserve wood is challenging. Nowadays, the wood used in most kinds of construction is either kept dry or protected using chemical preservatives. Heavy-duty toxins based on arsenic and mercury are banned, but their historical use is worth remembering when you refinish older wood. Most lumber used inside the structure of a building is not treated with chemicals, unless termites live nearby. Less toxic copper or zinc naphthenates are painted on or infused into window frames or other wood that might get damp. Wood touching the ground or used in decks is impregnated with solutions of alkaline copper quaternary (ACQ).[5] The priority otherwise is to keep lumber dry.

Wood is not the only material in a home that fungi like to eat, of course. For moulds, the built environment is a smorgasbord. In addition to wood fibers from lumber, there's human food, compost, cotton and linen from carpets and furniture, cellulose in wallpaper or drywall, skin cells and hairs that drop off human bodies, and droppings left behind by bugs and mice.

Damp Corners: Kitchens, Washrooms, and Leaky Pipes

The moist corners in a building are homes for moulds that grow on paint, silicone caulking, rubber gaskets, and just about anything with paper or cardboard in it. Water vapor billowing from sinks, kettles, and dishwashers condenses and drips from the coils of fridges and freezers. The warm water from dishwashers and washing machines soaks into the seals between the doors and cabinets. Those dark stains on rubber gaskets? They are often colonies of *Exophiala* species, one of the better-known examples of an informal group known as "black yeasts." Under a microscope their budding forms look like typical yeasts, but instead of making creamy colonies, they look more like dark tar or oil. They are not considered "true" yeasts because of this dark pigmentation and the fact that most of them also make hyphae. In nature, some cause sap stain in living trees, or diseases of fish. In houses, they form a grungy gray film in detergent dispensers. They tolerate the alkali ingredients in soaps that kill off most other moulds—they even eat some of the hydrocarbon additives that give detergents their oomph. Their ability to switch between hyphal and yeast-like growth allows them to adapt to the hot or cold, wet or dry environments of these appliances. A few *Exophiala* species grow at human body temperature, but the ones in dishwashers don't seem to cause disease. Even so, it's a good idea to wipe the gaskets of your appliances with soapy water or rubbing alcohol now and then, and clean food crumbs out of the filters.[6]

The caulking around the edges of sinks, toilets, bathtubs, and windows often turns pink or black from growth of another black yeast, *Aureobasidium pullulans*. Slimy fungi like *Aureobasidium* tend to stick to surfaces, and their spores don't find their way into the air easily. This species protects itself from drying by exuding the water-absorbing polysaccharide pullulan. The fungus makes a lot of this gelatinous slime when grown in a fermenter. Pullulan is used in hairsprays and skin creams, or polymerized into translucent films used to make dissolving breath-freshening strips. The wild communities of this fungus, which grow on stone, wood, and leaves, have distinctive DNA fingerprints in each geographical region, just like human populations do. But in the built environment, its genetic markers are all scrambled together, and the same clones seem to live in bathrooms just about everywhere. The theory is that the semidomesticated strains spread internationally, from home to hotel room to home, when tourists inadvertently carried its spores around on toothbrushes and hand soaps. *Aureobasidium*, like dry rot, seems to be urbanized.[7]

The efficient ventilation systems of most modern buildings prevent humidity from getting too high. Older buildings often have old humidifiers, dehumidifiers, or air conditioners—or all three—that use cooling coils to remove humidity, then drip the condensation into reservoirs. Moulds grow on the organic matter that accumulates in these pools, and fans circulate the spores throughout the ventilation system to join the hundreds of other kinds of spores floating around in the rest of the house.

The Desert Air: Xerophiles, Mites, and Mould Spores

The most conspicuous domestic reservoir of moulds is house dust. Most of the inside cavities of buildings are as warm and dry as a desert, and it's not always clear where the moulds are growing. The bits of dirt, sand, and fluff on floors and furniture and under our beds might seem mundane to most people, but they're surprisingly full of microbes. A sample from a vacuum cleaner is easy to collect and study. Although house dust is a vibrant ecosystem for dust mites and some fungi, a mixture of spores released in other parts of the building also ends up in the canister.

Scientists used to think that Western homes harbored about a hundred fungal species, with maybe forty or fifty in any one building, because that was how many grew when we sprinkled house dust on agar media in the laboratory. Now we know it is many more. In 2008, when new next-gen DNA barcoding methods became available, I decided to try DNA barcoding to look at the fungi in our house's central vac system. Most of the canister was filled by a cushion of dog fur. But at the bottom was 2 inches of dark gray powder so fine that the gentlest puff of air caused a mini tornado. I scooped a few tablespoons into sealed plastic bags and brought them into the lab. When the next-gen sequencing data came back, I was modestly surprised to learn that we share our house with something like six hundred fungal species. Among them are plant pathogens blown in from the yard, yeasts and moulds actively spoiling food, soil fungi tracked in on our shoes, and the expected "typical" house dust fungi. The clear signals of conifer endophytes were a puzzle until a picture of our Christmas tree slid across my computer's screen saver and reminded me of all the needles vacuumed off the floor. There was also the faint signal, just three sequences, of a poisonous *Amanita* species only ever seen in Japan. How did that get there? Despite all the fur, we didn't detect canine pathogens. Later studies of many other homes in several countries, using more refined survey methods, suggested that the mould concentration in our house was relatively normal.[8]

Next-gen techniques are so sensitive that it is often hard to be sure what the results mean. Do the rarest microbes, barely rising above the baseline, matter at all? Or do they have significant effects despite their low abundance, like a single pheromone molecule that can excite insects for miles around? Because DNA stays intact for a long time, it doesn't tell us whether the microbes that contained it were living, dead, or passing through on their way to somewhere else. It's like being in a cathedral and unable to distinguish the active worshippers from the tourists or dearly departed. The important fungi in buildings are the ones who live there, not the transients. Trying to figure this out from a dust sample is like trying to determine the ingredients of a recipe after it has been turned to compost. Filtering out the background noise of vagrant bacteria and fungi in microbiome analyses, and identifying microbes that are really living inside, takes some ingenuity.

If we want to remove fungal contamination from a house, we need to find out where the moulds actually hide. One amplifier is houseplants, which bring their own assortments of saprobic moulds, epiphytes, and rhizosphere fungi into our homes. Potting soil is often pasteurized to kill off fungal spores. Of course, there is a mould for that. Commonly known as the peat mould, *Chromelosporium fulvum* often survives the heat treatment. Because its usual competitors are killed off and no longer keep it in check, it lays a cinnamon-colored carpet of powdery spores across the whole soil surface. This can be a serious problem in greenhouses, which may find tons of soil suddenly overgrown. Through the microscope, the peat mould looks like a long slender hand with hundreds of tiny round spores popping out all over the fingers. If it gets too enthusiastic, it hoovers up so many of the loose soil nutrients that delicate plants like African violets start to suffer. When weakened, these juicy-leaved plants with their profuse flowers are treats for several moulds, including the noble rot mould *Botrytis cinerea*. *Botrytis* often blows over from the compost bucket and colonizes the soft wet leaves with its gray fuzz, mopping up any sugars lying around and leaving behind dry brown patches. It sends up branched sporulating structures that look like microscopic bunches of grapes. The same fungus causes a serious disease in vineyards, but rather than producing musty odors, some strains concentrate the sugars in the grapes used to make expensive Sauternes and Tokaji dessert wines.

Dead leaves on plants, indoors or out, are an open invitation for black moulds. They sporulate happily on dead herbs and grass outside, but also inside on the dead brown stem and leaf tissues of plants that are watered too much or not enough. Under a microscope, you often see the club-shaped spores of *Alternaria alternata* and branching bead-like chains of *Cladosporium cladosporioides*. Although they are often blamed for allergies, these two moulds are so ubiquitous that their significance in indoor or outdoor air is difficult to evaluate. Outside, the abundance of their spores rises and falls in synchrony with the amount of decaying herbage in the spring and autumn. But inside, they grow year-round. All they really need is moisture from a little leak or condensation to get going. In one place I lived, there was always condensation on the ceramic floor of the poorly ventilated basement bathroom. *Cladosporium* grew in a faint haze across the tiles. Further, the upstairs shower drain leaked through to the drywall basement ceiling, enough to support expanding rings of *Alternaria* colonies. I wiped them up every now and then, but they always came back.

Among the most common fungi in homes is a distinct group of xerophilic moulds that spend their lives in dust and preserved food. Whether their spores blow in from outside or drift over from the compost bucket, they germinate in the arid dust. Their hyphae snake out through the detritus searching for digestible nuggets and a few molecules of water. In nearly every home of mine, the tiny yellow sexual structures and greenish asexual spores of *Aspergillus* species hid behind wall hangings, on clothing in closets, and on leather baseball gloves stored in humid basements. They don't produce much in the way of mycotoxins, but in warm, humid buildings (or leather-rich car interiors) they often flare up and can be quite allergenic.

Another house dust xerophile, *Wallemia sebi*, is one of the few basidios that grow like a mould. It's kind of the domestic dog of the fungal world—brown, fuzzy, and seems to like living with people. It apparently took up residence with humans quite recently, perhaps brought inside hidden in table salt harvested from tidal flats, or in ultra-sweet liquids like honey or maple syrup. *Wallemia* often makes a dramatic peak in next-gen sequence analyses of house dust, but we were hardly aware of it previously because it grows so slowly on normal agar media. We need to add five times as much sugar and replace a fifth of the water with glycerol to make the medium xerophilic enough for it to grow. But when we do that, we find it in almost every house. One of its sneaky habits is nibbling on chocolate, which is a good hiding place for a mould with dark brown spores. Otherwise it is just a nuisance, although its tiny spores can provoke allergies.[9]

Several xerophilic moulds interact with tiny relatives of spiders called dust mites. In nature, dust mites hide in bird and rodent nests. They may have entered human dwellings riding on the backs of mice and discovered the pleasant desert environment awaiting them. Two species now scramble around in the sediments of most human dwellings, the "European" (*Dermatophagoides pteronyssinus*) and the "American" dust mite (*D. farinae*); despite their geographical names, both are widespread. They are almost invisible to the naked eye, but under a scanning electron microscope they look like mechanical monsters from science fiction films: squat and tank-like with protracted, bent legs and carapaces covered with long, spiky hairs. They live as long as three months and drop a few eggs every day.

Dust mites are also xerophiles. The wall-to-wall carpets of Western homes trap spore-laden dirt deep within their piles, and it finds its way into cushions and other textiles too. The moisture from our breath and sweat is enough to get dust mites eating, mating, and defecating. Awake or asleep, humans constantly shed a blizzard of dead skin flakes that mites like to chew on, so our bedding is a hot spot. After two years, 10 percent of a pillow might be the droppings and shed exoskeletons of millions of mites. Not only do mites eat our dead skin, they also snack on fungal hyphae and spores, and on textile fibers softened by mould growth. Some of these spores pass through the mites' guts and tumble out at the far end; then they germinate and the moulds grow on the mite droppings. Some scientists consider this relationship a casual symbiosis. Although mites and moulds seem to enjoy each other's company, they often live apart.

The Air We Breathe (Part 1): Mould Particles, Mycotoxins, Allergies, and Asthma

Since the early 1980s, we have become more concerned about the potential health consequences of airborne moulds in buildings. Parents of students sitting in "portable" classrooms built in the parking lots of overcrowded schools began to report that their children came home smelling of mould. Drawing a line between mould growth in a building and medical effects on an occupant can be a delicate process.

We automatically associate environmental allergies and asthma with reactions to pollen from outside, but mould spores, mites, and dried pet saliva or dandruff can all trigger symptoms. Humans evolved in a world covered with moulds, and most of us cope well with normal levels. Because more fungal spores float around outdoors in the spring and autumn, some mould allergies are seasonal. But if moulds amplify in a house, allergies can persist all year and may get worse. We each react to different moulds in our own way. Although individual mould species or metabolites sometimes cause specific concerns, the overall concentrations of fungal spores and MVOCs correlate with the health of any built environment. With so many species present in any building, pinning down the culprits that irritate a specific person is difficult. Standard skin-prick tests for mould allergies tend to focus on species that are more common outdoors, rather than evaluating antigens from *Wallemia* or *Aspergillus*, which are more likely to amplify inside.

Particles of frayed dead hyphae and spores chewed up by mites are potent components of airborne dust. Whether the mycelium in these fungi is alive or dead, as much as 60 percent of its cell walls consist of a polysaccharide called glucan.[10] When glucan is inhaled, it causes inflammation and elevated body temperature. This leads to the intense fever known as a cytokine storm, which is typical of many respiratory infections. Glucan is a major trigger of asthma symptoms for the 20 percent of patients hypersensitive to moulds. Despite its importance in asthma, glucan sensitivities are not yet part of routine testing. Some people are quite allergic to mite feces, and 30 to 40 percent of asthma cases in North America are related to dust mites. Buying hypoallergenic bedding and replacing your pillows every year or two can help lessen exposure in that most intimate corner of your home.[11]

The most notorious of the dark moulds, *Stachybotrys chartarum*, is often called stachy or the toxic black mould. In the mid-1990s, *S. chartarum* was implicated as a cause of fatal lung hemorrhages in about a dozen babies living in poorly maintained older wooden homes in Cleveland, Ohio.[12] In nature, stachy grows on dead plant material that has a lot of cellulose, like straw. In homes, it forms black splotches on wallpaper, ceiling tiles, insulation, or drywall (also known as wallboard or Gyproc), which absorbs water and swells when it gets wet, leaving the outer layer of paper to go mouldy. When stachy is alive, its spores are sticky and don't easily get into the air. So they rarely appear in air samples, although dried fragments of dead spores and mycelium do get airborne. In a wall cavity or under the floor, stachy can spread over several square yards without anyone noticing. Its spores have high concentrations of a potent mycotoxin called satratoxin. If inhaled, satratoxin and glucan from the cell walls inhibit the amoeba-like white blood cells (macrophages) in our immune systems, leading to serious inflammation and bleeding. Black mould colonies don't always mean stachy, though. Usually they are colonies of the comparatively benign *Alternaria* or *Cladosporium*, which should be cleaned up but are unlikely to cause more than a few sneezes.

I'm often asked what people can do in their own homes to reduce risks from mould growth. For most people and for most buildings, simply opening the windows, changing the bedding, and moving the compost along will stop resident fungi from amplifying and resident humans from wheezing and sneezing too much. If you do have sensitivities to dust, replace the carpets with tile or wood flooring. Vacuuming regularly and installing a furnace fan with an effective high-efficiency particulate air (HEPA) filter will help remove allergens from the air rather than recirculating them throughout the house.[13] You can treat mild mould allergies with over-the-counter antihistamines. But if you have asthma, sensitivities to environmental allergens, or a compromised immune system, consult your doctor to see if more intensive hygiene methods or medications are needed. And if your home is flooded as the result of a natural disaster, the mould growth is unlikely to be subtle, and informed help to deal with it should be readily available.

If in the course of daily life you do find small mould colonies on walls or ceiling tiles, clean them with solutions intended specifically for this purpose and approved by bodies like the American Industrial Hygiene Association. Don't use bleach: it dries into crystals that irritate the lungs and could be more troublesome than the fungus you are trying to get rid of. If you have visible colonies of stachy or other species, or ongoing problems, hire a qualified remediation expert. Make sure your contractor has experience with moulds—many focus on other indoor problems like asbestos. Qualified workers will wear appropriate personal protective equipment and follow local building codes when cutting out and replacing the contaminated material. And be sure they find and repair water leaks or sites of condensation, or you will end up in the same situation again.[14] Sometimes the source of moisture isn't obvious. A colleague of mine tracked a minor mould problem in a school to a specific section of the music room where trumpeters and trombonists from the school orchestra were emptying their spit valves onto the carpet.

As long as we live in homes and spend time indoors, we will interact daily with the microbial world. We can now detect with more and more accuracy which fungi live with us, and we can try to manage our interactions with them. Some of these species are indicators of problems and others cohabit with us peacefully. What's less clear is whether any of them may actually be helpful in buildings. In the early part of my career, my research focused on finding biological controls for sap stain and decay. We didn't get very far, in part because the lumber industry was then busy fighting to keep chemical fungicides registered, and in part because we just didn't understand enough about the ecology of fungi in wood to make biocontrol work. Some building engineers wonder whether it might be possible to add harmless or beneficial "probiotic" fungi to buildings during construction or to design structures that encourage such fungi, in order to inhibit harmful moulds or decay fungi. These musings are usually met with skepticism, as unconventional ideas usually are.

The reality is that our buildings—like trees, crops, and humans—are mortal. Over time, saprobic fungi will turn all of our built environments into compost. And there's one kind of building where none of us want to end up—the hospital. Our efforts to conceal ourselves from symbionts and pathogens eventually catch up with us. Like all the other environments we've considered for their fungal dimensions, our bodies are ecosystems too. We are all tied into nature, from beginning to end.

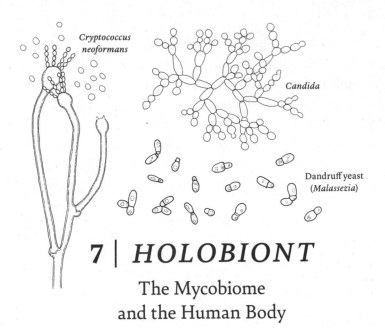

Cryptococcus neoformans

Candida

Dandruff yeast
(*Malassezia*)

7 | *HOLOBIONT*

The Mycobiome
and the Human Body

IN THE PAST TWO DECADES, DNA sequencing has let us look inside our human bodies in ways never before possible. With the Human Genome Project we gained a draft of all 3.2 billion nucleotide base pairs in our hereditary material.[1] We don't yet know the complete meaning of all the words that these letters spell, but the parts that we do understand illuminate some of the molecular mechanisms underlying human health and disease. It has shown us that one person's genome sequence is 99.9 percent identical to any other individual's. And sometimes a single nucleotide change in a single gene can cause disease. Another surprise is that about 25 percent of our DNA does not code for genes and that about 8 percent comes from viruses now permanently melded into our chromosomes.

More recently the Human Microbiome Project has surveyed the microbes on and in our bodies. Using the magnifying glasses of the polymerase chain reaction (PCR) and next-gen sequencing, scientists have detected thousands of times more symbiotic bacteria and hundreds of times more fungi than anyone had ever guessed. This constellation of micro-organisms, known as the human microbiome, varies considerably from one person to the next, from body part to body part, from hour to hour, and from infancy through old age. We've seen complex multiorganismal symbioses like this before, between trees, their ectomycorrhizal and endophyte symbionts, and rhizosphere fungi. Now we know that we are like that too. Almost every living thing big enough to see—wild and domesticated animals, reptiles, fish, and all our friends and relatives—is a collection of symbionts. Among animals, only some bats and insects, in particular some ants, seem to lack a prodigious bacterial microbiome. As we saw in chapter 3, this kind of mega-multipartnered symbiotic association is known as a holobiont. Even most fungi have bacterial symbionts living inside them or latching on to their outer walls. Russian matryoshka nesting dolls must have been inspired by microbiology. Humans are walking ecosystems.[2]

In mammal holobionts like us, microbial symbionts do seem to be heavily skewed towards bacteria. We have about 37.2 trillion of our own cells and something like 20,000 to 25,000 genes on our chromosomes. A slightly higher number of symbiotic bacterial cells are in our bodies (earlier estimates of ten times as many are now discounted), and they contribute an extra 50,000 genes to keep our digestion and other internal functions advancing and adapting.[3] But the contributions from fungal genes tend to be downplayed. Only about 0.01 to 0.1 percent of genes in human fecal samples are from fungi. One reason for this low number may be a medical preoccupation with bacteria, and the use of detection methods that often overlook fungi. But it may also just reflect a reality that fungal symbionts are less frequent in mammals than they are in plants or trees because bacteria prefer liquids and fungi prefer solids. Bacteria do better than most fungi at our body temperature of 98 degrees Fahrenheit and readily adapt to the lower oxygen concentrations in our digestive systems.

The inside of the human body is warmer than the average temperature in any outdoor ecosystem on Earth. The so-called fungal infection–mammalian selection (FIMS) hypothesis suggests mammals rose to dominance because dinosaurs were killed off by fungal infections. After the Chicxulub asteroid hit the Earth and our planet began to cool, dinosaurs were unable to increase their body temperature enough to combat pathogenic moulds, whereas mammals had evolved a self-regulating thermal system. What we call fever—an increase in body temperature—protects us from disease. Even today, the closest relatives of dinosaurs, birds, suffer more fungal diseases than mammals do. Infectious bacteria therefore cause more human diseases than fungi do, and we know comparatively little about the fungal component of our microbiome—our mycobiome. Still, DNA surveys have identified about four hundred species in the human mycobiome, and that number keeps increasing.[4]

Microbes colonize our bodies from our diet and from surrounding environments. Until the Human Microbiome Project, we mostly considered them aggressive, infectious pathogens that swoop in from outside and attack. That idea underlies the germ theory of modern medicine, which aims to wipe out microbes with antibiotics and other treatments. But we are beginning to understand that only a tiny proportion of the millions of bacteria and fungi in our environment are harmful, and that a healthy microbiome naturally balances the competing interests of individual microbes. More and more we see microbes colonizing our bodies and holding each other in check to ensure an "everybody wins if we work together" outcome.

Microorganisms on and in our bodies stimulate or suppress our immune system. Our immune system includes several kinds of amoeba-like cells collectively known as white blood cells that engulf and kill unwelcome viruses and microbes. When the body detects an invader, white blood cells travel to the site of attack and slide through the walls of veins and arteries into the affected tissues of the body. Like bouncers in a nightclub, they expel the undesirables and let the beautiful microbes in, at least when everything is working properly.

A few fungi grow on us or in us and seem to be truly symbiotic partners. We've been aware of some of them for a while, in particular those on our skin, but their contributions to our health—positive or negative—are a puzzle. Who are these fungal teammates? Which ones should we fear, and which should we welcome as essential or neutral partners?

Skin: Dandruff and Other Fungal Discomforts

Our skin is the surface that interacts with the outer world. When we say that someone has beautiful skin, we are really admiring how gracefully their bacterial and fungal partners interact with their epidermis. We tend to disdain exuberant microbial growth and the associated odors on our scalps and in the humid folds of our crotches, armpits, or toes. But the list of fungi detected by next-gen surveys of our skin is quite extensive. It reads a lot like the lists of moulds we find living in house dust, which suggests that we pick up and drop off most of them in places in our homes. Among the most prevalent are dandruff yeasts, which make an almost continuous coating over our skin. Although we've known about them for more than a hundred years, nobody thought very much about these fungi until scientists started to study the microbiome. Then we found out that about seven billion people are colonized by only a few species of basidiomycetes in the genus *Malassezia*. Different species of dandruff yeasts occupy different regions of the body. Some are more common on other mammals and move back and forth when people handle pets or farm animals.

Up to ten million yeast cells are found on the average scalp, each of them living about a month. What are they doing there? The layer of yeasty paste seems to both protect our skin from drying out and discourage potentially infectious microbes. Dandruff yeasts lack the genes to make their own lipids, so they soak up fats from their hosts. If you want to culture them, you need to add olive oil or milk fats to the agar. The flakes that most of us associate with dandruff are simply the result of our epidermal cells ejecting to make sure yeast cells don't smother our skin. A normal adult sheds about thirty to forty thousand dead skin cells per day, or up to 9 pounds per year.[5] Normally this is enough to keep the dandruff mycobiome and associated bacteria of the skin under control. But the human-*Malassezia* symbiosis seems a bit unsteady, and there is a tipping point when dandruff yeasts become a pest and the skin cells last only a few days before being banished. Applying a glutinous shampoo laced with selenium sulfide, a chemical that represses the growth of dandruff yeasts, usually tames them again.

Doctors only worry about dandruff yeasts when they irritate skin in a serious way; for example, when they cause scratchy, dry, flaky patches of eczema. Severe dandruff can develop into conditions needing medical treatment, like seborrheic dermatitis, which causes itchiness and irritation on warm, oily areas of the body. In tropical climates, tourists sometimes suffer from discolored patches of skin called tinea versicolor (also known as pityriasis versicolor or sun fungus). And when *Malassezia* gets into the wrong tissues it can provoke a dramatic immune response. This is dangerous for premature babies receiving lipids through catheters or for immunocompromised adults being given intravenous nutrients.[6] Dandruff yeasts are probably commensal or slightly mutualistic symbionts that drift towards parasitism if they have a chance.

How significant are fungi on skin to our health? Hyphae are seldom seen when the epidermis is examined under a microscope, so it is yeast cells and scatterings of spores that are normally present. *Saccharomyces* shows up a lot. So do moulds that make a lot of spores, like *Aspergillus* and *Penicillium*. Not only do the fungi from your home and outdoor environment leave their traces on you, but by molting all those dead skin cells colonized by dandruff yeast, you in turn shed a fungal fingerprint onto your surroundings.

Although dandruff is usually benign, other skin infections are more contagious, including athlete's foot, ringworm (caused by a fungus, not a worm), and the uncomfortably named jock itch or crotch rot.[7] Family doctors are very familiar with these common skin conditions. About 70 percent of people worldwide have fungal foot infections in their lifetime, and I am one of them. Although by most criteria I would not be considered sporty, every few years a warm itch breaks out between my pinky and "ring" toes, the typical starting point for athlete's foot. It feels as if someone has wiped the web with heating lotion, then scratched at it with a scalpel. The irritated skin blushes red, and the outer layer softens into loose, grayish-green, rubbery flakes that rub off when scratched. The fungus suppresses the immune response of the skin long enough for hyphae to colonize the outer epidermis. The problem is that the hyphae then grow into the nail bed. And the toenail is so rugged and impenetrable that antifungal drug molecules can't get through to the hyphae. The infection seems impossible to shake.

Over the years I've tried several different over-the-counter antibiotic creams, rubbing them between my toes twice a day for a few weeks. Clotrimazole seems to do the trick for me, but doctors can prescribe different classes of drugs or enroll your toes in drug trials, or you can buy exotic laser treatments or miracle cures off the internet. The success rate of all these interventions remains about the same, which is to say that the condition often lies dormant for a while before it flares up again. My symptoms recur every few years, usually in winter when my feet are wrapped in thick, warm socks in heavy, humid boots. Perhaps I should feel some professional pride for hosting a pet fungus on my feet, or at least a commensal that becomes a parasite now and then. It's likely to stick with me for life, and to become more aggressive when I am older and my immune system is on the decline.

My feet are likely hosting an invasive species. Most athlete's foot, jock itch, and ringworm infections are caused by one fungus, *Trichophyton rubrum*, which originated in Africa but probably hitched a ride to other continents in the nineteenth century when people started wearing closed shoes. *Trichophyton rubrum* doesn't reproduce sexually; instead, it seems to be a clone. And once a few hyphae install themselves in the outer layer of skin, the tiny spores of the fungus easily get tangled into towels, socks, or other clothing, giving them a chance to jump to fresh, hairless skin in public changerooms. This species and the various others that cause skin infections are close relatives of moulds that live in soil and favor the pelts, hooves, and antlers from deceased animals lying about there. Perhaps these treats were abundant enough to favor moulds that could make protease enzymes to break down the keratin and collagen that make up hair, nails, and skin. Many have given up entirely on plants and their cellulose as a food source.

Dandruff yeasts and other skin fungi rely on us dropping or swapping epidermis cells so they can get around. Maybe some of our social customs, like shaking hands, hugging, kissing, and the rest of it, are subtle manipulations by the members of our mycobiome so they can find their way to new homes or find themselves new mates. But if a fungus wants to participate in the inner social network of the human holobiont, it needs to find a way in.

The Gastrointestinal Tract: *Candida*

Much of the diversity of the human microbiome is inside the body. Our skin has regular pores and gaps, but the most obvious entrance for a microbe is through our mouths. Our digestive system, which comprises the throat, stomach, small and large intestines, and other organs, is a lot like the fermenters used in biotechnology—or, if you prefer, fermenters are like our gut. Everything is temperature and pH controlled. Nutrients and liquids shuttle from one compartment to the next so that larger molecules can be broken down into smaller pieces and reassembled in more useful forms. With all its pockets, side branches, twists, and turns, the digestive system is a sophisticated symbiotic pipeline optimized to digest complex foods. Most of the symbiotic bacteria in our microbiome live in our gut. For some parts of the digestion and waste management process, bacteria completely take over. Our own genes and enzymes need not deal with those steps at all.

Insect guts are the traditional hangout of wild yeasts, and several species find their way into our innards too. One of the common fungi inside our digestive tract is our old buddy the brewer's yeast. It floats around in the intestines of about half of adults, presumably picked up from handling dough or drinking unpasteurized beer—or perhaps by kissing trees. We're uncertain whether it's a true symbiont because its cells tend to pass through us in a few days. It is usually considered harmless, although some forms of inflammatory bowel or autoimmune disease correlate with elevated levels of antibodies against brewer's yeast. A particular strain, sometimes given the name *Saccharomyces boulardii*, was discovered after French microbiologist Henri Boulard noticed native Asians chewing on lychee or mangosteen skins as a natural remedy for cholera. It is sold as a probiotic to ease stomach and gut problems like ulcers, and especially to remedy diarrhea in people with AIDS or discomfort associated with infections by the anaerobic fecal bacterium *Clostridioides difficile* (usually known as *C. diff*).[8]

But it is the yeast-like *Candida* species that are the most frequent and enigmatic fungal symbionts in the human digestive tract, and in the vicinity of its various entrances and exits. *Candida* species live with about half of us, especially in our mouths, stomachs, intestines, and vaginas. Two closely related species, *Candida albicans* and *C. tropicalis*, are common; there are several other rarer species. They are frequently called dimorphic, which means they grow in two forms: as yeasts in some conditions and as hyphae (or chains of swollen cells called *pseudohyphae*) in others. The yeasts are nicely streamlined vessels for sailing along in moist tissues. Their floppy hyphae tend to get tangled up and are unsuitable for navigating through wet entrails. In *Candida*, the growth forms change in response to different temperatures or oxygen concentrations.

The presence of *Candida* on and in our bodies is a mixed blessing. We can still only speculate about the precise contributions of these yeasts, but they are so common in healthy people that there must be some benefits. Some microbiome studies show that the bacteria and fungi in the gut keep each other under control by marking off territory with noxious metabolites, as if we have two internal police forces each keeping watch over the other. Further, at least in mice, *Candida* primes the immune system to be more reactive against infections by other fungi. It also seems to form antibiotic barriers around vulnerable portals into the body, such as the urethra and anus. This action may block other intestinal symbionts from dripping into areas where they are unwelcome.[9] *Candida*, however, is known more for its pathogenic than its beneficial properties.

When humans are born, *Candida* spreads from the mother's birth canal and colonizes the baby's gastrointestinal tract by seeping in through the mouth. Within a few months, it settles into much of the infant's gut. In the child's first year, *Candida* is a major cause of diaper rash, but then the immune system learns to tame it. As we mature, *Candida* sometimes gets out of control, probably because of changes in the bacterial microbiome, and blossoms into the infections known as candidiasis. White pustules form on the mucous membranes of mouths or vaginas. From a distance, these resemble the patchy feather patterns of a particular bird, hence its common name, thrush. *Candida* often contaminates hospital equipment and is very difficult to clean up. Tainted catheters can result in deep tissue and blood infections in immunocompromised patients. Candidiasis then becomes one of the most lethal fungal infections, with a mortality rate higher than 40 percent.

The Air We Breathe (Part 2): *Aspergillus* and Other Fungi Growing in Lungs

Every time we breathe in, the rush of air sucks spores, dust particles, pollen, and pollution past the stiff hairs and moist membranes of our nostrils, through the trachea, and into smaller and smaller branches and branchlets of the respiratory system. Floppy little hairs swat at the intruders, trying to bat them out, but some evade the obstructions and crash, cushioned, into the alveoli of the lungs. These tiny round sacs, just the right size for fungal spores, are where oxygen and carbon dioxide pass between the air and our blood. The immune system of healthy lungs drenches intrusive spores with white blood cells. We expel the mucus left behind by coughing, sneezing, or snorting. But for spores that can hide in the smallest niches, or have tricks to outsmart the immune system, lungs are a great place to be. There is oxygen, a steady supply of food leaching through from the blood, and high humidity. As long as they can manage at our body temperature, the spores can germinate and explore.

Many of the truly life-threatening fungal diseases start out in our lungs. At first, many mimic viral or bacterial infections—coughs, chest pain, fevers, pneumonia—and are often misdiagnosed as a cold or the flu. On X-rays, the shadowy, shapeless blobs are sometimes confused with tuberculosis or cancer. And if the prescribed treatments target the wrong disease, the fungus continues to grow. If it remains untreated for long enough, it finds its way out of the lungs and spreads through the body.

The classic fungal respiratory diseases are caused by a trio of related ascomycetes. They are known as histoplasmosis (histo, or Caver's disease, caused by *Histoplasma capsulatum*), blastomycosis (blasto, or Gilchrist's disease, caused by *Blastomyces dermatitidis*), and coccidioidomycosis (coccidio, or valley fever, caused by *Coccidioides immitis*).[10] All three are endemic to parts of North America, but histo in particular is an invasive pathogen on several continents, perhaps spread by migrating bats. Fortunately, the infections are not transferred between people. They begin when people inhale spores from soil or from masses of bird or bat droppings. Instead of making more hyphae, the spores of histo and blasto respond to the warm, liquid environment and start to bud into small round yeast cells. Histo yeasts are engulfed by the white blood cells but not killed; those of blasto are too big to be consumed. They flow through the blood but also accumulate in the lungs. The inhaled spores of coccidio don't turn into yeasts but swell into sac-like cells called spherules. These burst and release hundreds of new spores, each repeating the cycle until the lungs clog up with fungal cells. Thanks to fungal drugs, these diseases are seldom fatal if accurately diagnosed. But people with AIDS, or those whose immune systems are repressed for other reasons, often suffer severe consequences from coccidio or histo infections.

Nearly all DNA surveys of human lungs report *Aspergillus fumigatus*, but no one would ever call it a symbiont. Its spores are almost everywhere, so it is no surprise that we breathe them in. In nature, this mould is a saprobe that breaks down plant debris. It amplifies dramatically in warm conditions, like self-heating compost piles, where its spores can be so abundant that they hover in a haze.[11] It seems to get stirred up by excavations for building foundations or during renovations, and then its spores get drawn into buildings. Surveys in hospitals find low levels of *A. fumigatus* spores behind false ceilings, in ventilation systems, and in other nooks where moulds are prone to hide indoors. It also likes potted plants and is a contaminant in some marijuana-growing operations. Fortunately, it is easy to recognize when it grows on agar media because of the long columns of dry spores that jut up from its colonies like grayish-blue towers of cigarette ash. Individual spores of *A. fumigatus* are the perfect size to get lodged in the alveoli. For most people, however, inhaling a few spores provokes a response that is similar to ordinary hay fever, and for those with a robust immune system, the usual antihistamines will suffice. The spores get mopped up by the white blood cells, and the fungus causes no further problems.

Heavy exposure to the spores of *A. fumigatus* can lead to hypersensitivity, though, just as some people develop severe allergies with their second bee sting. People who over-react to these spores suffer the constricted airways, coughing, wheezing, and breathing difficulties that characterize serious asthma. Severe reactions may cascade into anaphylactic shock. Prolonged exposures to *A. fumigatus* lead to a condition called allergic bronchopulmonary aspergillosis (ABPA), in which the mould grows along the lining of the bronchial airways, resulting in inflammation and sometimes permanent lung damage.[12]

Here the line between allergy and infection gets fuzzy. *A. fumigatus* is a thermophile, quite happy growing at human body temperature, and if the spores settle into the alveoli for long enough they germinate and hyphae start to grow. They begin to release a mycotoxin called gliotoxin, which represses the immune system. The mycelium coils up like wads of cotton, growing into masses called aspergillomas that lodge in the lungs and sometimes penetrate other nearby tissues. The disease, a type of chronic pulmonary aspergillosis, is a major hospital-acquired infection and kills thousands of people every year. Patients with deep skin punctures called "major barrier breaks," or those with a suppressed immune system, are particularly at risk. If the fungus spreads to the blood or central nervous system, the disease is about 95 percent fatal. A small percentage of people who survive serious flu or SARS-CoV-2 infections (COVID-19) later develop *Aspergillus* infections—which should encourage you to keep up with your vaccines.

AIDS and the Rise of Fungal Diseases
Until recently, humans had little fear of dying from a fungal infection. Human immunodeficiency virus (HIV) changed that. HIV causes acquired immunodeficiency syndrome (AIDS), a condition in which the body's immune system is too weak to stop infections that an uninfected person would easily shake off. When the AIDS epidemic hit in the 1980s, doctors were caught off guard by the sudden increase in virulent fungal diseases. Previously rare or benign diseases became more common or deadly because the weakened immune system couldn't repel them. The catalog of fungi that might turn up in hospital patients rose from about a hundred to more than five hundred. After antiretroviral treatments against HIV were introduced in the mid to late 1990s, the annual global number of deaths from AIDS fell from about 2 million

to less than 1 million, with the developing world lagging behind. In sub-Saharan Africa alone, between 15 and 25 million people still live with AIDS. Between 45 and 50 percent of deaths among patients with AIDS are caused by fungal infections running amok because of reduced immune function.[13]

One of the most serious diseases to co-occur with AIDS was the previously rare *Pneumocystis* pneumonia (PCP), an infection that was often the first clinical symptom. The pathogen, a single-celled fungus with lemon-shaped cells, colonizes lung tissue. All mammals have *Pneumocystis* species in their lungs. They seem to be ancient symbionts that joined their evolutionary destiny to mammalian lungs long ago and are seen nowhere else. The species associated with humans is *Pneumocystis jirovecii*, although it was often called *P. carinii*; the latter species name is now used for a pathogen of dogs.[14] Children often test positive for *Pneumocystis*, but only 20 percent of adults do. There are usually no symptoms and it is eventually cleared out by the immune system. It becomes an opportunistic pathogen only when the immune system is suppressed. Nevertheless, *Pneumocystis* cells can transmit from host to host through the air, and asymptomatic humans are considered the main source of infection in people with AIDS. Diagnosis and treatment have improved, and mortality rates have fallen from as high as 40 percent to 10 to 20 percent, but there are still about 10,000 cases per year in the United States. Where does *Pneumocystis* fall in the symbiotic continuum? Rather than an invasive pathogen, it often looks more like a commensal that becomes a parasite when the immune system fails to keep it in check.

Today the most serious fungal disease of people with AIDS is cryptococcosis, or crypto. Around the world every year, about 220,000 people are infected by yeast-like basidiomycetes called *Cryptococcus*.[15] About 180,000 die annually, some 165,000 of them in sub-Saharan Africa and most of them with HIV. Roughly the same number of people with HIV die every year from tuberculosis. In nature, crypto grows as a yeast in soil, in bird droppings, and on tree bark and produces hyphae only in anticipation of sex. Infections start when the resulting sexual spores are inhaled. The yeast form has a mesh of fibrous polysaccharides that swells out as a halo-like capsule. In the wild, this capsule protects the cells from being digested when they are swallowed by predatory amoebae; in the lungs, it protects them from similar attacks from white blood cells.

Patients who develop cryptococcosis progress from coughs, fatigue, headaches, and fever to weight loss, chest pain, vomiting, and stiff joints. Like most other fungal diseases that start in the lungs, it usually does not spread from one patient to another. Once inside a human host, the two species involved behave differently. The cells of *C. neoformans* cross into the central nervous system, forming cysts and lesions in the brain. This leads to a meningitis that is 100 percent fatal if untreated, and still 30 percent fatal after treatment with antifungal drugs. Until recently, the closely related *C. gattii* was considered a tropical curiosity growing on *Eucalyptus* trees, until it was noticed infecting porpoises and dogs in North America in the late 1990s. Then it was recognized as an overlooked cause of crypto in people with AIDS. Its infections tend to remain in the lungs, where it causes pneumonia. As with PCP, most people with intact immune systems who are exposed to crypto don't become ill.

At the same time fungal diseases were rising because of reduced immune function in people with AIDS, they were becoming more common in patients whose immune systems were compromised for other reasons. The awareness of PCP in people with AIDS led doctors to recognize that the same disease often occurred after organ transplants or chemotherapy for cancer. Now 60 percent of PCP infections are diagnosed in people who don't have AIDS.

The Mycobiome and Antifungal Drugs

Presently, at least a billion people are infected by pathogenic fungi of one kind or another. The annual death toll is between 1.2 and 2 million people per year, more than double that from malaria and in the same ballpark as the 1.4 million deaths from tuberculosis.[16] When it comes to designing or discovering effective drugs to treat fungal infections, it is unfortunate that humans are so closely related to fungi. What is toxic to them is often also toxic to us. Antifungal drugs can be tough on humans, and if an infection is not life-threatening, the cure sometimes feels worse than the disease.

The first effective antifungal antibiotics were discovered during the bioprospecting boom of the 1950s to the 1970s. Amphotericin B (AmB) and nystatins were both isolated from two species of actinomycete bacteria in the genus *Streptomyces*.[17] Both antibiotics are polyenes, long lipid-like chains that bind with ergosterol, a structural component of fungal cell membranes. This makes the cells fatally leaky. Because ergosterol is chemically similar to cholesterol in animal cell membranes, polyenes are somewhat toxic to humans and can cause intestinal problems, fever, fluctuations in blood pressure, and in the long term, kidney failure. Today, most fungal infections are treated with chemically synthesized antibiotics known as triazoles, which have bouncy names such as fluconazole, itraconazole, clotrimazole, or ketoconazole. These also interact with ergosterol but interfere with enzymes involved in its biosynthesis, meaning the fungal cell membranes cannot form. Again, some patients react badly to these drugs, and about 1 percent of patients using triazoles suffer gastric discomfort, muscle and joint pain, headaches, or ringing ears.

Our growing discomfort with the use of antibiotics is partly because they affect the stability of our microbiomes. Humans are already unstable holobionts, with a vacillating population of bacterial and fungal species. As with any ecosystem, microbial succession takes place in our bodies as we age. For example, a fungus like *Candida* behaves differently at different times of our lives, and it can be a dangerous pathogen in one part of our body and a commensal in another. These variations make treatment difficult. When we attack one pathogen with an antibiotic, we attack the whole microbiome. Opportunistic former friends take advantage of the changed conditions and transform into serious parasites. For example, in Crohn's disease (a type of inflammatory bowel disease), the acid-tolerant yeast *Debaryomyces hansenii*, commonly found in fermented cheeses or meats, appears to colonize the gut lining after its symbiotic bacteria are disrupted by antibiotic therapy. The immune response to the yeast cells causes inflammation that interferes with the normal healing of wounds in the bowel. Many chronic disorders that are difficult to manage or cure seem to correlate with changes in our microbiomes. Conditions like obesity, heart problems, and autoimmune disorders (type 1 diabetes, lupus, rheumatoid arthritis), or the many diseases that lack satisfactory genetic explanations, may develop after our mutualist symbionts are knocked off-kilter, a situation called *dysbiosis*.[18]

Medicine is evolving to include the view that disease can be an ecological problem. Instead of believing that every microbe is a pathogen and eradicating them all, our strategy will shift to figuring out how to coax the wavering loyalties of opportunistic symbionts back into equilibrium where we need them. The same will be true for dealing with the problems we have with fungi in nature and in our homes. We need to embrace biological complexity if we are going to be wise stewards of our world. The health of our microbes is our physical and mental health; their genes are our genes. They are part of our inheritance from our parents, partners, children, houses, foods, and everywhere we've ever been. If we are going to make peace with fungi, we need to be aware of their biodiversity and embrace their talents for biodegradation, symbiosis, and biochemistry that make them such significant players in the environment. Only then will we be able to work with them effectively for our own prosperity and health, while they also collaborate with us.

PART 3

THE MYCELIAL REVOLUTION

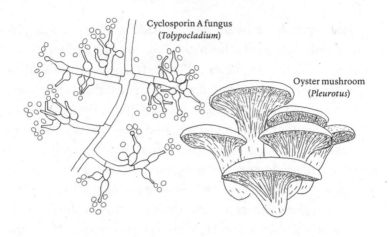

Cyclosporin A fungus
(*Tolypocladium*)

Oyster mushroom
(*Pleurotus*)

8 | *MYCOTECHNOLOGY*
Fungi for the People

WHETHER HUMAN CIVILIZATION was initiated by tribes of agricultural apes or apes interested in brewing larger vats of beer, it emerged because of our discovery of biotechnology, the adaptation of biological systems or living organisms for our own purposes. "Biotechnology" is a modern term, but we have been planting forests, growing crops, raising animals, and producing alcohol, bread, and other fermented products for thousands of years. Fungal symbionts, saprobes, and pathogens all played their part in the successes or failures of these enterprises. So far, we've exploited just a few species. But because fungi have such an array of talents, and there are thousands of other species that could be considered, innovators are increasingly experimenting to adopt more of them for new purposes. Four characteristics of fungi make them particularly attractive collaborators, spawning possibilities for new mycotechnologies.[1]

First, tubular hyphae and spreading mycelia are different building blocks from the bag- or brick-like cells of plants, animals, or bacteria. Their fibrous nature means they can be reoriented into durable threads, flexible textiles, or spongy bundles. Mycelium spreads like a liquid made of fibers, surrounds other materials, and glues them into three-dimensional composite solids. When appropriately stabilized, these products resist being pulled apart, torn, bent, or crushed. Processed mycelium is a valuable addition to the toolboxes of engineers and designers as they rethink building materials and textiles. How can we use mycelium to make strong, malleable structures? What new, previously impossible products could we design with such building blocks?

Second, fungi make bucketloads of metabolites that affect other living cells, both good and bad. So far, we've focused on mycotoxins, antibiotics, alcohols, flavor and odor compounds, and signaling molecules like pheromones. Fungal organic acids are well established in industrial processes and the food industry. As medical researchers look to exploit knowledge from the Human Genome Project and new understanding of our immune systems and microbiomes, what new classes of drugs might arise from fungal metabolites? Can other metabolites replace by-products of the fossil fuel industry?

Third, fungi are great decomposers. A lot of our past efforts went towards preventing rot and decay. But our current focus on recycling, reducing waste, and remediating damaged ecosystems puts fungal biodegradation into a happier light. Fungal enzymes like cellulase, ligninase, and amylase break complex polysaccharides into sugars that are the feedstock for biofuels. Others replace corrosive chemicals such as bleach in detergents. Can fungi be used to degrade plastics and reduce the footprint of landfills, or to address pollution caused by accidental spills?

Finally, fungi have a great track record as creative collaborators. Without symbiotic mycorrhizae and endophytes, our forests and farms would be shadows of their present selves. Yeasts, moulds, and mushrooms provide us with food and vitamins. Can foresters and farmers better use endophytes and mycorrhizae to reduce plant disease, protect against invasive species, and reduce pesticide and fertilizer use?

But managing living organisms is complicated. We need to be cautious, because we don't want exuberant fungal participants to produce toxins or spread into places we'd rather they not be. If we improve our understanding of fungal biology and biochemistry, and interactions of fungi with other organisms in nature, we increase our chances for success in using them for our own purposes. And they may help us out in ways we have yet to imagine.

Industrial Mycotechnology: Organic Acids, Plastics, and Enzymes

The use of fungi in industrial microbiology started in 1919, about the same time that the word "biotechnology" was coined. The first applications exploited their talents at producing chemicals and enzymes. *Aspergillus niger* was the first hyphal fungus to be used in an industrial process, and it is a very comfortable collaborator in the human world. It sits near the top of many surveys of moulds found in soil, in compost, in homes, or on human skin. Its sooty black heads throw out thousands of spiky asexual spores that germinate quickly wherever they land. Its hyphae get a jump start on sleepier moulds, which makes it a serious nuisance as a lab contaminant. But *A. niger* is a talented cellular factory that pumps out streams of diverse metabolites and enzymes. It thrives in the liquids of fermenter tanks, like those invented for breweries.

After World War I, the American pharmaceutical company Pfizer used *A. niger* to make citric acid. This is the compound that makes citrus fruit sour. The mould produced far more than the lemons used previously, and the fermentation was a lot faster than growing a fruit tree. Without the $2–$2.5 billion worth of citric acid manufactured each year, there would be no squeaky-tooth mouthfeel from cans of cola. It also binds to metal ions, making them more soluble for industrial use or for cleaning the calcium out of your coffee maker.[2]

Myco-entrepreneurs dream up new applications for moulds constantly. For example, itaconic acid and similar organic acids are fermented on an industrial scale by *Aspergillus terreus* and then linked together to make acrylic plastics, latex paints, and many other products. Along with *Aspergillus niger* and its relatives (like the kōji mould), brewer's yeast and species of *Trichoderma* are the main sources of industrially important fungal enzymes. Both in nature and in fermentation tanks, they are excreted from growing hyphae and do their work outside of the living cell. They can be purified and used to drive industrial processes without the need for a living mould. A lot of industrial processes happen above 100 degrees Fahrenheit, and the enzymes that are used need to work at those temperatures. It makes sense that the thermophilic moulds of hot environments, like compost heaps or wood chip piles, would make heat-stable enzymes. Enthusiastic ads for laundry detergent frequently tout the power of their fat-degrading lipase and protein-breaking proteinases without mentioning that many come from thermophilic moulds like *Thermomyces lanuginosus*. The denim in "stone-washed" blue jeans is usually softened with cellulases made by the mould *Trichoderma reesei*. Presumably calling them "fungus-rotted jeans" is a marketing no-no.

The more we discover about fungi, the more we realize that we need to use caution. After a century of industrial use, the recent discovery that some cultures of *A. niger* make ochratoxin A was a surprise. This kidney toxin is usually associated with other *Aspergillus* and *Penicillium* species in stored grain, coffee, wine, and beer. Most industrial strains of *A. niger* don't make it, but this revelation was a valuable reminder to check for hazardous by-products when moulds are grown in high volumes or used in new processes.

Fungal biotechnology advanced and diversified as the twentieth century progressed. What began as a search for organic acids and enzymes broadened when the metabolic creativity of fungi became clearer. Medical demands for new drugs broadened the exploration to a wider range of fungi.

Bioprospecting: Mining for Drugs

The discovery in September 1928 of the first antibiotic, penicillin, is iconic. Alexander Fleming (1881–1955) became one of the most famous scientists of the twentieth century, and his breakthrough is often used to illustrate the importance of "accidents" or serendipity in science. Fleming studied lysozyme, an enzyme in human tears that is our natural protection against bacterial eye infections. His laboratory on the one-and-a-halfth floor of London's St. Mary's Hospital near Paddington Station was basically a closet with three tall, sealed windows. He crammed his microscope onto the lab bench among racks of stagnant bacterial cultures and bottles of noxious solvents. Returning from a family holiday,

Fleming pulled an old culture of a pathogenic skin bacterium called *Staphylococcus aureus* (known as staph) from a tower of glass Petri dishes. A dusty green colony of a species of a mould contaminant had crept in from the side. In front of the spreading mould (today known to be one of the most common indoor xerophiles, *Penicillium rubens*),[3] there was just clear agar. Where the cream-colored bacteria should have been, there were only ghosts of dissolving colonies. Something was killing the bacteria.

Fleming did what any of us would do. He took the mouldy Petri dish to morning break and showed it to his colleagues over coffee and a cigarette. Newspapers claimed that one of his colleagues nibbled a chunk of the mould and said it tasted like Stilton. A year later, he described his observations and experiments in a short paper, famous for its vagueness and lack of detail.[4] He named the bacterium-killing substance penicillin.

Nobody paid much attention, and Fleming went on to other things. When war flared up a decade later, a team of Oxford chemists led by the irritable, square-jawed Australian Harold Florey (1898–1968) and his collaborator Ernst Chain (1906–1979) adopted penicillin. They worked out the molecular structure and purified enough to try out as a drug. Eventually they had enough to inject into infected mice; the rodents recovered. Then the team treated an Oxford policeman who was suffering from a persistent infection from a rose thorn. The penicillin had to be repurified from his urine and reused. A little bit was lost with each passage. His recovery continued until the supply depleted, and then he died.

As the Blitz intensified, the team worried that the Oxford facilities might be bombed. They painted spores of the *Penicillium* culture onto their lapels before heading home each night, so that they could re-isolate the fungus if the lab were destroyed. To keep the potential miracle drug away from the Nazis, Norman Heatley (1911–2004), the lab's creative jack-of-all-trades, transferred the research effort over the Atlantic to Peoria, Illinois. They needed much more penicillin before they could use it as a drug. Mary Hunt (1910–1991), an assistant in the Peoria lab, routinely bought rotting fruit at the local farmers market, earning herself the nickname Mouldy Mary. A new strain she isolated from a collapsing cantaloupe produced hundreds of times more penicillin. During World War II, the North American supply of penicillin was less than half an ounce, about the weight of a pencil. A decade later, it was mass-produced and one dose cost about the same as a glass of milk. The same strain, mutated to further increase its output, is still used to produce the drug today.

Fleming, Florey, and Chain won the 1945 Nobel Prize in Physiology or Medicine. Fleming's fame remains controversial, especially for those disappointed that Heatley did not share in the award. Though he was a shy man with a stiff, awkward smile, Fleming nevertheless enjoyed attention. He repeated his mouldy story to eager interviewers for decades. And he autographed dried agar colonies of his fungus to serve as presents for friends, colleagues, and royalty. A sample recently sold at auction for £11,863 (roughly US$16,000), an impressive sum for a dried-up house dust fungus. One charm of his discovery is that every student of mycology or microbiology sees the same antibiotic phenomenon in contaminated experiments. Its significance is now obvious but went unappreciated until Fleming thought it through.[5]

Penicillin launched the age of antibiotics and a golden age of drug discovery. In the 1950s, pharmaceutical companies often asked employees to collect soil during their vacations, a practice that came to be known as bioprospecting. Armed with teaspoons, they traveled the world and scooped samples into 35 mm film canisters or plastic baggies. Back in the lab, technicians diluted the dust and spread it across agar culture plates in the hope that new moulds would grow and pump out novel metabolites. The chemicals were separated from the cells and purified. Lab scientists designed panels of experimental tests, called bioassays, to see whether the isolated compounds had interesting biological effects. They screened metabolites from a huge diversity of microbes, animals, and plants from every conceivable marine, terrestrial, tropical, polar, or extreme environment. Fungi and actinomycete bacteria were highly productive. The successes were dramatic. Streptomycin, an antibiotic isolated from the actinomycete *Streptomyces griseus*, cured tuberculosis, gonorrhea, and syphilis, adding to the miraculous mystique of antibiotics.[6]

Taking a metabolite from discovery to drug is an expensive, high-risk investment. The competition among researchers and rival pharmaceutical companies was intense. In a kind of medical arms race, most discoveries were protected with patents or kept as trade secrets.

One race was to find drugs to assist with human organ transplants. In 1967, the South African surgeon Christiaan Barnard (1922–2001) performed the first human heart transplant. The patient lived eighteen days before dying of pneumonia. Successful transplants remained a dream for a long time. It wasn't just the physical trauma of surgery that was dangerous. White blood cells in our immune system interpret organs from other bodies as alien tissue and attack them. Scientists at Sandoz, the Swiss pharmaceutical firm where Albert Hofmann synthesized LSD, hoped to solve this problem. They recovered strains of a slow-growing white mould called *Toly-pocladium inflatum*, the asexual form of a zombie fungus, from soil from Norway and the United States. The strains made a complex circular molecule that they named cyclosporin A. In nature, the seepage of the compound confounds competing moulds. It doesn't kill them but makes their hyphae branch over and over again. Instead of growing longer, their hyphae tie the colonies into tight knots that can't spread. In human blood, the drug interrupts T cells (a type of white blood cell) before they attack the transplanted organs.[7]

Cyclosporin saved about half a million transplant patients in the first thirty years of its use, earning royalties of about a billion dollars per year for its discoverers. It is also used for other immunological diseases, like rheumatoid arthritis and psoriasis. About five thousand people have heart transplants each year, and organ transplants in general are now considered routine. Unfortunately, cyclosporin has some side effects, especially on kidneys, so it needs to be monitored carefully. And fungal diseases are very common in recovering patients. The repression of the immune system by cyclosporin has a similar effect to HIV infections in AIDS. It reduces the body's normal protections against infection. It's ironic that one of the most celebrated fungal products of the drug-prospecting era also increases the rate of infection by dangerous pathogens like *Aspergillus fumigatus*.

Most of us grew up when the fear of infectious diseases was subsiding. We tend to forget what a revolution antibiotics were and how recently this happened: the first person saved by penicillin died in 1999. The average human life span increased by about fifteen years after World War II, partly because of these new drugs.

Unfortunately, overuse can lead to antibiotic resistance in bacterial and fungal pathogens. This resistance happens because one dose of the drug kills many but not all pathogen cells. Some withstand the first attack. Survivors able to break down or disable the drug pass the ability on to their descendants. Overexposure to antibiotics lets the rare resistant cells amplify after susceptible cells of the same species die. Antibiotics, like penicillin and streptomycin, that once protected us are losing their effectiveness, and defeated diseases are re-emerging. Multiple drug resistance is also emerging. For example, the wound pathogen *Candida auris*, discovered in Japan in just 2009, invades the bloodstream of hospital patients on several continents and is undeterred by most available antifungal drugs. The United States Centers for Disease Control and Prevention (CDC) reports that 2.8 million Americans are infected by antibiotic-resistant bacteria and fungi each year, resulting in more than 35,000 deaths. Both the number of infections and the number of deaths increase constantly, but statistics from different countries are difficult to compare.[8]

Surprisingly, fungicide use in agriculture is also leading to antibiotic resistance in human pathogens. We spray chemicals called azoles over field crops to prevent fungal diseases like smuts and rusts. When the fungicides drip onto the ground, most soil fungi die, but those resistant to the compound survive and multiply. These pesticides have a similar chemical structure to triazole drugs used to treat human fungal diseases, like candidiasis and aspergillosis. If azole-resistant strains from agricultural fields infect a person, they are not easily stopped with triazole drugs. Resistance to triazole drugs is exploding in many agricultural regions in human pathogens like *Aspergillus fumigatus*.[9]

Presently, many antibiotics and other drugs are produced by genetically modified organisms (GMOs), whether fungi or bacteria. The biotechnology boom of the 1980s followed the discovery of scissor- and glue-like enzymes that snipped genes from one species and spliced them into the chromosomes of another. Known as recombinant DNA, these techniques transplant genes from wild microbes into a host already trained to serve as a cellular factory in a fermenter. For example, the insulin my mother injected for her diabetes in the 1960s was not the human version but came from the pancreases of fetal pigs. Now, molecules identical to true human insulin are made by GMO cultures of brewer's yeast (or the bacterium *Escherichia coli*, known as *E. coli*) grown in fermenters. The microbes are altered to express the human gene, which produces proteins identical to the human form of insulin. Genetically modified brewer's yeast is also used to make the hepatitis B vaccine, eliminating side effects associated with the original vaccines that included blood products.[10]

Much more natural biological creativity exists beyond what we cut and paste into workhorse microbes. What surprises might we find if we open our eyes to the talents of the millions of other fungi hiding in nature?

Mycofoods and Biofuels

Mushrooms are cholesterol-free, low in calories and saturated fats, and good sources of amino acids, protein, fiber, minerals, vitamins, and antioxidants. In the 1980s, wondering if we could get the same nutritional bounty if we just ate mycelium, food engineers began looking for fungal cultures that could make lots of protein or lots of hyphae. They looked for fungi that grew well in fermenters. Some stayed in tight little balls, others expanded into cloudy orbs, and less inhibited strains filled the whole vessel with a chowder-like soup. Even if the results looked unappetizing and didn't taste like much, the engineers knew they could add texture and flavor later.

One popular product that resulted from these experiments is Quorn. To make this mycoprotein, the potato pathogen *Fusarium venenatum* is grown in fermentation tanks. After it is filtered, the mycelium is rearranged to give it a similar mouthfeel to chicken. This meat substitute was introduced to the market in 1985 and is still sold in Europe. It is only available in a few countries partly because it is based on a plant pathogen that sometimes makes mycotoxins.[11] Most fungal food entrepreneurs prefer to exploit mycelium of species with a history of safe use as food, like species of edible mushrooms.

Apart from their suitability for vegetarian diets, fermenter-grown fungal foods are environmentally friendly. The starting material is often agricultural waste, like wheat straw or maize stalks. This approach employs energy stored in the starch of crop residues that can't be used for human or animal food. Amylase enzymes extracted from *Aspergillus niger* or *A. oryzae* break down the starches, releasing sugars used to grow the edible mycelium. The process uses between a fifth and a tenth of the water needed to produce animal protein, and food is ready in days or weeks rather than months or years. Meat grown in Petri dishes from animal cell cultures gets more media attention, but the economics are likely to favor fungal protein in the long run.

Forest wastes could serve as a valuable feedstock for fungal fermentations too, but the cellulose in wood is too well protected by the lignin matrix. Decades of experiments with lignin-degrading enzymes from wood-decaying basidiomycetes have found ways to turn this polymer into useful phenolic chemicals, but large-scale industrial applications are still rare. Lignin is an important problem to solve. A lot of air pollution emitted by pulp and paper mills is a consequence of the need to separate lignin from cellulose by simmering the wood fiber with sodium sulfide. In Sudbury, we always knew the prevailing winds had shifted when the miasma of rotten eggs rolled in from the mill town of Espanola 40 miles away. Today, those facilities use enzymes like cellulases and ligninases from thermophilic fungi to assist with pulp and paper manufacture. Combining these biobleaching and biopulping processes with other odor-reducing strategies means that Espanola is no longer as fetid. But globally, we still use only 5 percent of the 100 million tons of lignin trapped in dead trees each year. The rest is incinerated.[12]

Bioethanol and other biofuels are important and steadily expanding alternative energy sources in developed countries, and they rely on fungi.[13] Usually built near farmland, the factories producing biofuels look like miniature oil refineries with tanks, pipes, and conveyer belts leading out of grain silos. The starting material is agricultural waste ground into fine meal: sugarcane bagasse, corncobs, soybean stalks, and rice straw. The raw mash is mixed with fungal cellulases that break the cellulose into sugar and hemicellulases that deal with the other polysaccharides. Alternatively, microscopic algae are grown in open-air ponds and photosynthesize sugars that can be used as feedstock. In either case, the sugars are fed to brewer's yeast in fermenters, and the resulting ethanol is distilled and added to gasoline. Most fuel sold for automobiles is 10 percent ethanol in Canada and the United States, and 5 or 10 percent in Europe. In Brazil it is 25 percent. Blends of gasoline with as much as 83 percent ethanol can power appropriately designed engines in flexible-fuel vehicles. Bioethanol delivers less energy per gallon than gasoline, but also less than half the carbon dioxide. Although mycotoxins from the original crop don't end up in the concentrated ethanol, they may sometimes remain in the waste solids, called distillers' grains. This is a problem if they are dried and used for animal feed.

The guts of cattle, horses, sheep, and wild animals like elephants, giraffes, and zebras have a mycobiome of mutualistic anaerobic fungi that belong to an obscure phylum with the Mary Poppins–like name Neocallimastigomycota. They are closely related to chytrids. Inside the animals, these fungi produce zoospores with either one or several whip-like flagella that wiggle through the gut and encyst between the tough straw fibers that the animals eat. The swelling and growth of their hypha-like cells bursts the tough plant tissue apart. The multiprotein cellulase complexes of species like *Neocallimastix frontalis* are so effective at breaking down cellulose fibers that they are nicknamed "cheese graters." These are the strongest cellulases known and are of great interest for breaking down the feedstock for bioethanol fermentation.[14]

Mycoremediation: A New Approach to Environmental Cleanup

Our present preoccupation with environmental issues often focuses on climate change and carbon emissions. Fungi are significant players in the global carbon cycle because as symbionts they support the fixing of carbon in plant tissues, and as saprobes their rotting activities release carbon dioxide. But pollution is an equally serious problem, and fungi will be key partners in addressing environmental toxins. Biopesticides and biofertilizers should make a significant contribution to reducing the use of pernicious synthetic pesticides and fertilizers. Fungi are also being mobilized to remediate contaminated wastes.

The use of fungi to detoxify or clean up oil spills, mining effluents, radioactive waste, and solid pollutants is called mycoremediation. Several moulds grow in hydrocarbon-rich or highly acidic environments and can use petroleum products as a carbon source. The absorptive tendencies of fungal mycelium have been tested to clean up after forest fires, nerve gas attacks, and fuel spills on land or at sea. For example, oil that washed up on San Francisco beaches after the 2007 *Cosco Busan* spill was mopped up with a mixture of human hair and oyster mushrooms. The disasters at the nuclear power plants at Chernobyl (Ukraine) in 1986 and Fukushima (Japan) in 2011 released radioactive fallout over large areas, including cesium 137, an isotope that causes malignant tumors and shortens life spans significantly. Concerns that Ukrainians eating wild mushrooms would suffer radiation poisoning led to the discovery that many fungi accumulate heavy metals (such as cesium 137) in their mycelium; similar contamination was noted in matsutake in Japan. But the tendency of mycelium to scrounge heavy metals suggests that mushrooms could be cultivated to extract isotopes from soil, with the mushrooms then harvested and incinerated to further concentrate radioactivity for eventual disposal. Later studies of the interior walls of water-cooling towers in the Chernobyl reactor also showed that several house moulds, like *Cladosporium cladosporioides*, absorb radioactivity in the melanin pigments that darken their cell walls. This mould was investigated on the International Space Station to see if it could be used as a self-replicating radiation shield in spacecraft. And the recently discovered yeast *Rhodotorula taiwanensis*, which seems happy to grow in highly radioactive environments contaminated with mercury and chromium compounds, could be useful for future mycoremediation operations in liquid acidic mining wastes.[15]

The fungi (and bacteria) that naturally colonize contaminated material can often process and break down toxins on their own. But, as with other kinds of compost, the succession of microbes varies from place to place and season to season and gives inconsistent results. With basic lab equipment, local scientists can test and formulate starter cultures based on mixtures of local species, which avoids the risks associated with non-native introductions. Because these treatments are mostly needed for garbage exposed to open air, they are more feasible for rural communities or remote facilities. So far, a lot of screening for mycoremediation has involved moulds like *Aspergillus* or *Penicillium*, a limited spectrum of thermophilic moulds, or mushrooms that happen to be on the shelf. The moulds grow quickly, like weeds, and that makes them easy to culture and lets them prosper on organic waste. But many such moulds produce mycotoxins or can grow at human body temperatures. In the short term, until a broader range of local species can be tested, species like brewer's yeast or *Aspergillus niger* that live almost everywhere and are generally recognized as safe would be preferable and less likely to cause health problems.[16]

Plastic pollution, the so-called plastisphere of packaging, water bottles, and abandoned toys, is a serious problem. Ascomycete moulds like *Aspergillus*, *Fusarium*, *Penicillium*, *Trichoderma*, and *Chaetomium* are often isolated from weathered plastic. Their hyphae lack the power to break plastic apart using physical force, but when the cultures are added to trash at the lab scale they produce enzymes that dissolve the plastics into more easily handled molecules. Those enzymes should be useful for reducing the 350 to 400 million tons of waste plastic—90 percent of it used just once—that ends up in landfills every year. Rather than adding fungi directly into garbage heaps, it may be that fungal enzymes will be added to plastic trash in liquid slurries.[17] A similar approach is being used to remediate soils contaminated by noxious chemicals like polychlorinated biphenyls (PCBs).

In tropical countries, local fungi are also being evaluated to treat the huge volumes of rubbish from coffee cultivation, 40 percent of which is waste biomass from the cherries. Often the fibrous, mucilaginous pulp is dumped into rivers, and caffeine and tannins leach out and poison fish. However, lab experiments show that composting the debris with moulds like *Aspergillus* or *Rhizopus* can remove 90 percent of the caffeine and 65 percent of the tannins. After that, the residue can be used as raw material for animal feed, biogas, or enzyme production or for growing smaller crops of edible mushrooms, especially oyster mushrooms—as long as that (so far unidentified) *Aspergillus* species isn't a pathogen and doesn't produce mycotoxins. In addition, filters made from bundles of dried coffee cherry husks seem to work to absorb toxic crystal violet in effluents from tropical dye factories.[18]

These examples suggest untapped potential for mycoremediation of land and water ecosystems at a scale affordable for developing economies. With their talents for biodegradation and biosynthesis of diverse metabolites and enzymes, fungi seem like a natural match for deliberate interventions to repair damaged ecosystems. The fact that only a few weedy moulds have so far been explored from the extravagant fungal diversity of tropical countries suggests that a significant opportunity for future innovation exists there.

Boutique Biotechnology

Since the turn of the new millennium, large-scale, industrial fungal biotechnology has scaled down into the entrepreneurial domain. The fungi have thrown in their lot with diverse citizen scientists, small start-up companies, designers, and artists. With new microelectronics and nanotechnologies at their fingertips, these impresarios are free from academic conservatism or the revenue expectations of the stockholders of large companies. Instead, they can focus on smaller niche markets. And like a hyphal mycorrhizal network, their creativity is spreading mycotechnology in so many different directions at once that to insiders it seems that a mycelial revolution is taking place.[19]

The most visible success stories are new twists on old beverages. Small breweries making artisanal craft beers are popping up everywhere in strip malls, industrial parks, and repurposed garages. Adventurous brewers are distancing themselves from the well-worn industrial starter cultures and experimenting with wild yeasts from flower nectar or insects. There are brews known as bumblebeer made from beetle belly yeasts, enhancing the usual malt and hops flavors that dominate industrial beer. Classic fermented brews are re-created from ancient yeast cells isolated from bottles retrieved from one- or two-hundred-year-old shipwrecks. And innovative fruity and sour beers are brewed with starter cultures based on yeasts like *Brettanomyces* or *Pichia* rather than the usual *Saccharomyces*. Social media is full of "likes" for the results, and happily there's no chance of mycotoxins because the true yeasts (class Saccharomycetes) don't make them.[20]

Similar experimentation is underway with cheese. Inspired by classic mould-fermented curds made with "accidentally" domesticated *Penicillium*, some cheese makers are trying to deliberately domesticate other species. Modern technology allows us to analyze the genes and secondary metabolites of various fungi, so artisanal producers try to enhance the aromatic metabolites and eliminate mycotoxins as they tame wild *Penicillium* strains. To optimize how their cheese looks, they pick segments that have white rather than colored spores. This can lead to complete and permanent white growth after only eight generations—it's almost as if the moulds realize they should hide.[21]

The leading edge of mycotechnology is the development of new solid materials made from mycelium.[22] The starters of choice are usually basidios like oyster mushrooms or the wood-decaying medicinal polypore *Ganoderma lucidum*, known as reishi. They are favored because their mycelium is unlikely to produce mycotoxins or provoke allergies (although high spore concentrations of oyster mushrooms can cause a hypersensitive response known as mushroom worker's lung). The mycelia of these mushrooms are grown in a liquid fermentation and then filtered out and processed. Or they might be spread over a large surface and then peeled off as sheets. The texture of a natural hyphal mass is usually like foam or cork, with the sponginess of a mushroom cap. But after harvest, mycelium is layered, kneaded, or rewoven into a diversity of consistencies, and then dried or killed with heat.

Hundreds of products are under development. Leather-like mycelial textiles with a realistic pebbled surface are being sold as fashion alternatives to animal skins or plastics made from petroleum products. Sheets of these materials can be milled to any size rather than being limited by the dimensions of an animal pelt, although one might wonder what they smell like if they get wet. Car manufacturers and electronics companies are injecting fungi into molds so that the mycelium will grow into specific shapes and can be used like Styrofoam, often as a packing material. Fungal composites are a blend of mycelium and other solids to make strong yet pliable materials that harden when compressed and treated with heat. The result can be as flexible or as rigid as needed, and these composites are replacing boards, bricks, and plastics. Using these building blocks, architects and artists are assembling walls, buildings, furniture, and even otherworldly art installations.[23]

These mycomaterials compete directly with traditional products and technologies. They require less energy and water, tend to be flame resistant, provide effective thermal or acoustic insulation, and are biodegradable. On an environmental spreadsheet, they outcompete materials made from nonrenewable minerals and petroleum derivatives as well as renewable resources like plant fibers or animal skins. If the cost and quality advantages are significant enough, mycomaterials should help us evolve to a more sustainable economy with a reduced carbon footprint. They will become mainstream rather than boutique.

They could also aim for the stars. NASA is considering myco-architecture for an eventual colony on Mars because hauling bricks, mortar, and wood from Earth would be extremely expensive. It would be cheaper to build small factories on the Red Planet, where mycelium could be grown and fashioned into building materials on the spot. The habitats of our first off-world colony may be grown and assembled from the mycelium of the first off-world fungi. Models show curved lightweight panels covering squat buildings like a turtle shell.[24] Like an artificial lichen, photosynthetic algae could be sandwiched between flexible layers of compressed mycelium, so that the buildings could harvest solar energy and produce their own oxygen.

A Fungal Fantasy

Set your imagination roaming, and try to envision what our lives and world might look like if we fully embraced our fungal friends...

You awake refreshed from a comfortable sleep, your lungs protected from the fungal allergens in your bedding by the hypoallergenic bedsheets on your mattress. As you lather up in the shower, your shampoo is optimized to keep dandruff yeasts under control. Afterwards, you apply a soothing facial moisturizer that includes homogenized mycelium of reishi, the "mushroom of immortality."

Like most humans, you disdain expensive farm-raised or fermenter-grown meats and enjoy a mainly vegetarian diet. Your breakfast flakes were produced from organic cereal crops optimized with fungal partners for pest resistance. For toast, why not try some artisanal sourdough bread made from a starter of exotic yeasts and bacteria and slathered with Marmite? As a savory, perhaps an imitation bacon created from compressed mycelium, or tempeh or a funky cheese crafted from newly domesticated moulds. If you need more protein to get your metabolism rolling, try the single-celled variety: a yeast extract or mycelial cake. It has abundant soluble fiber, a vibrant boost of B vitamins, and a customized flavor cocktail of microbial organic acids, aldehydes, and glutamates that satisfies your cravings for umami. Wash it all down with some yeasty kombucha or another fermented drink that is crafted to maintain the vigor and diversity of your gut microbiome. And your morning coffee? It is thoroughly tweaked by its fermentation with the healthiest yeasts, bacteria, and moulds to assure you experience a perfect blend of microbial aromas and flavors while naturally reducing caffeine to a level beneficial to your mood and health. You can rest assured that the waste coffee cherries were blended with indigenous moulds and bacteria into a nutritious compost to grow local edible mushrooms.

Refreshed by breakfast, you stretch and return to the bedroom to get dressed and head to work. No more pesticide-drenched, water-guzzling cotton for you! Fashioned from mycelium harvested from fermenters, your ever-so-soft garments are mostly shades of muted gray, green, and brown highlighted with explosive yellows, oranges, and reds inspired by the natural palette of lichens. The faux leather of your shoes is also compressed mycelium, which has been processed into a tough, self-repairing textile. Dressed and ready, you get into your new symbiotic car. It is partly powered by bioethanol distilled from agricultural waste products fermented by yeasts. The vehicle's electronic functions are powered by sustainable redox flow batteries made from mould mycelium, or ion-exchange batteries made from mushroom flesh. And the bumpers, side doors, and dashboard are all made of compressed and hardened fireproof mycoplastics.

Your office, like your house, is built from fireproof foundation blocks, wall panels, and interior surfaces all made from mycelium instead of wood harvested from forests or concrete made from nonrenewable aggregates. The soft cushion of your office chair, where you will spend the next eight hours, is a fungal foam that feels very much like sitting on a cushy mushroom cap. Any deliveries arrive perfectly protected by custom-grown mycelial cushions rather than Styrofoam. And your wall gardens are seeded with the appropriate mycorrhizal and endophytic fungi to ensure vibrant, psychologically supportive plants while retarding the growth of unwanted asthma-inducing moulds. The building's air circulation system filters out harmful fungal spores and hyphal fragments.

You and your family are healthier. Harmful moulds are banished from your accommodations. The symbionts in your holobiont body are tweaked and balanced so that *Candida* and the dandruff yeast remain focused on their essential duties and no longer cause you distress. If an infection arises, it is treated using microbiome-friendly antibiotics manufactured using brewer's yeast or the kōji mould as the cell-factory host. You manage stress and depression and reduce cravings for all the new delicious aromatic alcohol beverages (brewed using exotic yeasts) by using microdoses of psilocybin. These are just one part of your personalized psychomedical regimen, including customized blends of probiotics and prebiotics, carefully monitored by your nutritionist or doctor.

Bioluminescent lamps provide muted illumination for the roads and pathways of your neighborhood each night, keeping you safe while allowing birds and other wildlife to use moonlight for navigation. The light bulbs are miniature fermenters filled with the genes of fluorescent mushrooms transplanted into yeast. They glow perpetually with a soft light.

Overhead, the interplanetary spacecraft heads for Mars carrying small vials of fungal and algal spores. Extraterrestrial gardens will thrive from the mycorrhizae, endophytes, rhizosphere symbionts, and epiphytes as we extend our tradition of biological invasion beyond our own planet.

And your body? When you are done with all of this material splendor, you too are compostable. When your time comes, you will be recycled in a mushroom burial suit, ensuring a fashionable and environmentally friendly transition as your carbon rejoins the microbes.

This scenario may seem like fantasy, but none of these ideas are mine. These technologies either exist or are the subject of active research and commercial development.[25] Ideas grow a lot like hyphae do, as networks, making connections, mutating, spreading. Clever people are brainstorming, proposing increasingly wild ideas about what we might do with fungi. Some may come true; many won't. And when we consider the broader world—the needs of lower-income countries and natural ecosystems in peril—the promise of the mycelial revolution assumes a different face. Food production, drug development, antibiotic resistance, bioenergy, pollution, and waste management are all serious challenges that need serious attention. The unique properties and skills of fungi can help us out.

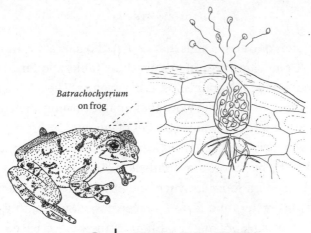

Batrachochytrium
on frog

9 | *THIRTY THOUSAND FEET*

Fungi and the Sustainable Planet

VIEWS FROM AN AIRPLANE don't immediately bring fungi to mind. After all, most of the hidden kingdom is microscopic or tucked away underground. But from an altitude of 30,000 feet above the surface of the Earth, evidence of fungal activity is everywhere. From there, you see the blocky outlines of commercial plantations that long ago replaced natural forests, vibrant from the activities of their endophytes and ectomycorrhizae. Disrupting the green canopies are irregular patches of gray, brown, or rust where trees attacked by bark beetles and the symbiotic fungi they carry are slowly dying. Above the flatlands, vast farms nourished by arbuscular mycorrhizae (AM) and rhizosphere fungi stretch out like squares on a checkerboard, except where native prairie hangs on in gullies or along the banks of meandering rivers. Sometimes you can see faded green crop circles that look like the landing sites of UFOs, a sign that crop pathogens are running amok.[1] Around agricultural towns, shiny tanks and ferment-ers sprout beside the grain silos and elevators. These are bio-fuel factories, where fungi work their magic turning plant and animal waste into renewable sources of energy.

A few specialized satellites remotely monitor crop and forest health by recording changes in the wavelengths of light reflected off foliage, which sometimes identify precise diseases.[2] Foresters and farmers receive weekly, monthly, or annual planting, spraying, and harvesting advice based on weather information, disease models, and the exact species or cultivars they grow combined with satellite observations of their land. It's kind of like "personalized medicine" for plants.

We have maps to illustrate continental drift, the migration of peoples, the flow of crops and livestock from continent to continent, and the lopsided distribution of prosperity. Maps of fungal migration, when we have them, trace the same paths. The view from above reminds us that all the living creatures of our world—including humans—are interconnected and depend on a healthy environment. We know this at an intui-tive level, but the precision and increased sensitivity of DNA barcoding surveys have brought the scope and intricacy of the biological world into sharp focus. We are learning more about our own genomes and microbiomes, microbes like fungi that we scarcely noticed before, and the complex relationships that drive ecosystems as varied as forests, farms, food, build-ings, and our bodies. Everything interacts and nothing acts alone. The better we understand how these systems work, the more problems we will be able to solve.

As we learn, we become more aware of the gradients between conflict and cooperation, and the tensions between the needs of various species, the environmental forces of the

Earth, and our economic activities. Fungi play a role in many of these situations, for good and for bad. To attain a sustainable world, we will need to balance competing interests. And in any realistic scenario, we have to remember that our experiences with fungi will not always be positive. Among collaborators will be opportunists. New fungal pathogens and epidemics will affect us and our oceans, forests, and farms. New fungal toxins will be discovered in foods and products that we previously thought were safe. As medicine continues to learn about the complexities and shortcomings of our immune systems, and the nuances of balancing our microbiomes, new allergies and hypersensitivities will emerge. How do we protect our own interests, and those of other living organisms, without disrupting Earth's balanced and interconnected ecosystems? How do we stall the so-called sixth extinction, where one species—us—wipes out thousands or millions of others?[3]

Symbiosis is a critical biological phenomenon and plays a big role in our interactions with fungi. Most forests are human-designed ecosystems where the symbioses are still intact and continue to be discovered and investigated. The significance of symbioses on farms was discovered only recently. Because we were unaware of them, AM symbioses in crop plants are much weaker than they are in native grasslands, and the endophytes that should protect leaves from insects seem to have disappeared. If these symbioses can be restored to their original state, or even strengthened beyond that, it will reduce chemical use. But the improvements can only be done in steps. In some cases we may need to go back to the beginning and re-create domesticated crops, being sure to carry along the symbionts this time. In the built environment, the ecology is almost completely accidental and symbioses are few. We now know we are symbionts ourselves, but our preoccupation with disease and the naive use of medicines seem to have thrown our microbiomes off balance. The real hope is that we come to understand that just about every ecosystem is built from communities of symbionts, and that every living thing has a microbiome. Now, when we construct artificial ecosystems, we can take these partners into consideration. These symbioses are critical to our health and to environmental health, and this realization may motivate us to actively restore the balance.

There are more questions than answers. But our progress highlights promising ways that researchers, citizen scientists, and policy makers are rethinking our relationships with fungi and proposing thoughtful solutions.

Policy and Regulations: Balancing Trade and Biodiversity
The migration of species has always happened, but it used to be that almost everyone stayed more or less where they started out. Countries and continents were discrete territories with their own assortments of creatures, plants, and microbes. Migration usually occurred at the velocity of plate tectonics; or the speed of a swimming spore or a drifting plume of spores, pollen, or flies; or occasionally faster when a hurricane, tsunami, or volcanic eruption stirred things up. In recent centuries, the global melting pot—the diaspora of people and their technology—gradually erased the geographical barriers that used to keep everything in place. Contemporary trade continues the tradition of deliberately transporting plants, animals, microbes, and people between continents. It is as if the planet is in the process of reblending itself into one giant ecosystem, the way it was 200 to 300 million years ago in the time of the supercontinent Pangaea. The process is

sometimes wistfully referred to as Pangaeafication.[4] Humanity is a big part of this process: invasion is in the eye of the beholder, and the biggest invader of all is us.[5]

Trade brings many benefits but also many problems. The movement of plants, animals, and microbes across oceans can overcome millennia of geographical isolation. Green lumber crosses oceans to be used for buildings, furniture, pulp and paper, and many other products we use to prop up our day-to-day lives. Wheat (from the Middle East), maize (South America), and rice (Asia) now grow almost everywhere, and much of the world's human population depends on these crops. With these raw materials come fungi, either in and on the wood and grains, or hitching a ride with other microbes, insects, or rodents interested in the same commodities.

Diseases and pests don't respect political boundaries, and our forests and farms often feel the effects. Quarantine agencies actively monitor borders, trying to prevent or slow down the passage of the most economically and ecologically threatening species—those on their "most unwanted" lists.[6] Monitoring allows us to detect diseases before they become established, and before foresters, farmers, or consumers are aware of them. Most of this scouting still happens on the ground. Experienced plant pathogen detectives learn the patterns of healthy and ailing crops and the typical field symptoms of disease. Quite often these surveys fulfill their purpose, and quarantined species are detected before they have a chance to start an epidemic.

As an example, outbreaks of potato wart, a disease caused by the chytrid *Synchytrium endobioticum*, have occurred in Europe and North America for at least a century.[7] The disease is spread in infected potatoes or by thick-walled microscopic cysts that survive in farm soils for up to thirty years.

So, shipments of potatoes are checked by inspectors whenever they cross borders. Infected tubers develop galls that swell into thick black cankers and are easy to spot. As well, to keep potato wart from spreading from its focal point in Newfoundland to other potato-producing parts of North America, the undercarriages of all vehicles leaving the island by ferry must be sprayed with disinfectant. Despite the care to contain potato wart, an outbreak on Prince Edward Island (PEI) in 2000 halted all potato exports from Canada to the United States, which did not have the disease. No DNA diagnostic test was then available, so inspectors used microscopes to search for cysts in thousands of soil samples collected from every square mile of PEI farmland. Meanwhile, my colleagues generated DNA sequences from potatoes involved in the outbreak and compared them to dried specimens kept as reference material from previous infections. The DNA from infected PEI potatoes was a match for a tuber from an outbreak of potato wart in 1908.

To develop a diagnostic test, molecular biologists evaluate DNA sequences from several genes. The scientific community agrees which are most useful as a barcode to identify the target fungus. Then they compare sequences from different fungi to tease out signature sequences that only occur in the target species. Our culture collection also held archived strains of other *Synchytrium* species that cause diseases on wild plants and weeds. After comparing all these barcodes, we found unique sections of the potato wart sequence and used them to develop a PCR test (see chapter 1) to detect and identify the fungus. This more sensitive diagnostic tool made surveying the soils much faster, and potato shipments waiting for export were quickly checked. The Americans accepted the results of these tests and reopened the border, and Canadian farmers sighed with relief as regular trade resumed.

Quarantine only works when we know what to look for. Lots of microorganisms slip under the radar, most of them harmless. But half the planet's frog species have declined since 1980, and about a hundred species of frogs, toads, and salamanders have gone extinct, largely due to chytridiomycosis. This fungal disease is caused by *Batrachochytrium dendrobatidis*, or Bd. Its zoospores encyst on the outer skin of the frogs but don't penetrate; nevertheless, the disease is about 90 percent fatal. The pandemic (or panzootic) apparently emerged from Asia in the early twentieth century and was probably spread initially by frogs hiding in pools of water in unexpected corners of machinery, like mining equipment. Then it spread widely with the deliberate trade of frogs as pets, for food, or for use in human pregnancy tests. Currently the DNA of Bd is found on the skin of about half the world's frogs, not only in waters close to urban areas and national parks but also in wilderness ponds and remote lakes, where the species can also survive as a saprobe. The current epidemic (one of several events lumped together as the "amphibian apocalypse") may be a consequence of the evolution of more aggressive strains, or perhaps cold-water amphibians are more prone to infection when exposed to climate change and pollution.[8]

Some pet frogs have been treated with the antifungal drugs terbinafine or itraconazole, otherwise used for human skin infections, but such applications are impractical outdoors. However, microbiologists discovered that the few frogs naturally immune to Bd often have a bacterium named *Janthinobacterium lividum* (Jliv) as a part of their skin microbiome.

Letting susceptible frogs swim around in water spiked with Jliv wins some of them immunity, but not all.[9] Managing native bullfrog populations and putting tighter reins on frog farming and export may be our best bets to slow the spread.

In 2013, a second fungal species, *Batrachochytrium salamandrivorans*, or Bsal, was found attacking the dramatic orange and black European fire salamander.[10] Originally from Southeast Asia, this disease seems also to have spread to Europe on imported pets. So far, Bsal has not been found in North America, home to 40 percent of the world's known salamander species. After loud croaking from activists sensitized by the slow international regulatory response to Bd, the United States Fish and Wildlife Service implemented a temporary ban on the importation of two hundred salamander species as pets in 2016. For once, the regulatory process acted faster than zoospores could swim.

Until recently we've been reluctant to discuss the risks of modern travel and world trade. Epidemics and pandemics of people, animals, and cultivated plants make us realize we have to do better. When neighbors cooperate, vigilance often pays off to stop invasions. (Unfortunately, some countries use allegations of infections and threats of quarantine as bargaining chips to lower prices on imported products.) We also understand better how and where fungi move across borders. In the past, we relied on expert taxonomists who knew what to look for. Now, the regular use of next-gen sequencing that allows us to survey DNA barcodes changes the game. Using these huge datasets, we can plot maps that show the true distribution of various fungi all over the world. Once we have a DNA barcode for what we think is a rare fungus, we can check archived next-gen data to see if it actually is rare. Our assessments of the risks of fungal disease transmission can be based on robust evidence.

Biological Collections: Balancing Scientific Research and Sustainability

For centuries, scientists removed plants and insects and rocks and fossils from their natural habitats to study and preserve them. Naturalists like Alexander von Humboldt (1769–1859), Charles Darwin, and Alfred Russel Wallace (1823–1913) explored the globe gathering specimens for their collections. Taxonomists in particular tend to be hoarders. We stuff as many species as we can find into our satchels or backpacks as we explore outside our own territories or on international expeditions. Some dried collections, called fungaria, are hundreds of years old and contain millions of specimens sorted into customized archival folders, bottles, or containers and preserved in climate-controlled cabinets.[11] Cultures of fungi, too, are isolated and saved for future observations or use in experiments. Microbial culture collections contain tens of thousands of living strains. Vials of spores and mycelium are also kept in suspended animation by freeze-drying or by freezing in liquid nitrogen at about minus 300 degrees Fahrenheit.

Improved DNA sequencing technology has suddenly made dead archival specimens of plants, fungi, and insects more important. If these reference samples were carefully stored, we can amplify and study their genes using new methods developed for ancient or fossil DNA. For fungi, we can access two centuries of the genetic legacy of invasive diseases, symbionts, and species making interesting metabolites.

Naturalists and taxonomists are no longer the only scientists interested in these specimens. Governments use them to resolve quarantine issues, as we did when dealing with potato wart. Scientists working for pharmaceutical companies and other enterprises collect fungal specimens hoping to develop proprietary drugs, new foods, or exotic mycelial building materials. But the commercial potential of fungi—historical and contemporary—has raised new concerns about biopiracy and the ethics of removing specimens from their natural habitat, whether for research or for profit. The genes recovered from the dead specimens are not just barcoding genes, but also functional genes. They can be spliced into workhorse lab strains to make genetically modified organisms that produce enzymes or metabolites that have never been studied before, and that may have commercial potential.

Fungi, especially microscopic ones, are an unopened Pandora's box in most countries. The millions of fungal species in nature are critical components of our natural heritage, and the tropics are home to most of them. Traditionally, researchers from Europe and North America simply took what they wanted and often then tried to sell it, or a product made from it, back to the countries where the material came from. In an effort to protect biodiversity and the property rights of nations, an international treaty called the Convention on Biological Diversity (CBD) was established at the 1992 Rio Earth Summit of the United Nations and extended by the 2010 action plan, the Nagoya Protocol on Access and Benefit-sharing.[12] These agreements oblige bioprospectors—explorers who collect biodiversity for any reason—to obtain written permission from the country, Indigenous Peoples, or landowners where they collect specimens or samples and to negotiate an agreement to share any profits.[13] Any effort to develop new drugs or other products from any kind of organism must now conform with these laws.

In addition to protecting species, the goal of these agreements is to support lower-income countries in the development of biotechnologies based on local microbes, and to include them as full financial partners in any products developed elsewhere from their biodiversity. The vision is that profits from future drug and industrial successes will support education and the development of local biotechnology. This will take decades, and the interim period is a heavy burden for economies that lack resources to support modern science. Until recently, taxonomic expertise and biological collections were concentrated in Europe and North America. But several programs to train students from lower-income countries in prestigious scientific institutes have changed the landscape.

As a start, individual countries are encouraged to establish their own culture collections. Cultures of indigenous microbes provide research material and the potential for innovation. For rare fungi or those in danger of extinction in nature, cultures offer the possibility to reintroduce species to the wild. So far no one has tried this at a large scale, but researchers in Finland reintroduced seven locally extinct wood decay fungi into spruce forests using cultures isolated in nearby localities.[14] Because of the risk of unwanted spread, the release of living microorganisms is carefully regulated in most countries. Usually only microbes that occur naturally in a region can be deliberately let go. For example, plant breeders often assess disease resistance in new crop cultivars by spraying spores harvested from Petri dishes or fermenters onto their test plots. The cultures have to come from the same area. Most strains in collections are used only in laboratory experiments, and if they are grown in large-scale fermenters, containment measures are enforced to ensure they don't accidentally escape.

Like the Svalbard Global Seed Vault or the proposed all-life lunar ark, our culture collections—or "Guardians of the Microbial Galaxy"[15]—preserve the genetic diversity of today's fungi for the benefit of unborn generations. Under the CBD, countries have the responsibility to study and preserve their own indigenous organisms, which gives them further incentive to survey, preserve, and control the distribution of their own resources, including fungi. All inhabited continents now host at least one major fungal culture collection and fungarium.

Previous efforts to stimulate economic development focused on exporting so-called appropriate technology from richer to poorer nations. Foreign aid organizations promoted crops, technologies, pesticides, and fertilizers that worked well in temperate countries but gave ambiguous results in the tropics, as seen in the Green Revolution. Now, biotechnology and mycotechnology are framed by international initiatives with labels like "bioeconomy" or "circular economy," which promote the development of sustainable, small-scale industries based on local organisms. The few species mycological entrepreneurs have explored are already producing value-added mycotechnologies. With the enormous fungal diversity indigenous to tropical countries, a broad spectrum of local symbiotic, saprobic, and biochemical fungal biodiversity waits in the wings to stimulate economic growth and encourage innovation. The key is to enhance rather than degrade local environments while developing robust economies.[16]

Citizen Science: A New Relationship With Fungi

Without realizing it, some people already know quite a lot about fungi. Orchid growers, for example, quickly learn about mycorrhizae because without them they would have no orchids. Out of necessity, gardeners learn about plant diseases and how to avoid them. People who make bread get interested in sourdough starters. Home brewers experiment with different strains of yeast. Anyone with access to a computer, a mobile phone and its camera, and an internet connection can take advantage of many of the scientific advances that have let us peer deeper into the microbial world. For a few hundred dollars you can purchase a pretty nice microscope with a built-in camera and start to document the microbes in your neighborhood. You can even send specimens to commercial services and buy a DNA barcode. Contributing fresh observations and data to science is no longer restricted to people with PhDs in industry and academia—we can all take part.

Increasingly, citizen scientists—regular, everyday people in cities and rural communities around the globe—are documenting the biodiversity of their own surroundings and sharing it on digital platforms like the iNaturalist website and mobile app. You can upload photographs (with the GPS coordinates, time, and date embedded in the image files), and the artificial intelligence (AI) algorithms will suggest the identity of the organism in the picture. The iNat community of specialists, both amateur and expert, checks the results. After two people have confirmed the identification, it is considered "research grade" and a dot is added to the international map of that species' distribution. At the moment the system works best for larger animals and plants and some groups of insects. Interest is growing and identifications of larger fungi, like mushrooms, lichens, and polypores, are improving. Special projects, such as time-limited bioblitzes (like the Continental MycoBlitz) or ongoing surveys, unite local, national, or international naturalists with common interests, building a solid sense of community. When I travel, I use the app as a portable field guide, and at home, I lurk in the background and try to help people identify their moulds.

In recent years, information about historical specimens and cultures—previously only accessible to specialists through card catalogs and institutional databases—was put online at the behest of the CBD. The Global Biodiversity Information Facility (GBIF) database collates the catalogs of the world's biological collections and complements the information collected by iNaturalist and similar citizen science initiatives.[17] Combined, they show the distribution of species as verified by professional researchers and as observed by citizen scientists. If a species seems like it might occur in a particular area but we have no specimens or cultures, iNaturalist can mobilize local participants to try to fill the gap. This allows dedicated amateurs, who often become experts without formal academic training, to contribute specimens to public collections and work with taxonomists who can access state-of-the-art technology. New observations and specimens accumulate at an increasing rate, along with knowledge about where species live and what plants or animals they associate with. This enhances molecular detections made in DNA-based surveys. If potential impacts on biodiversity are well understood, they can be balanced against economic concerns when political decisions, such as imposing quarantines, are being made. And if you like, you can look to see when and where in your area you might hunt for a particular fungus—chanterelles, for example.

It's an exciting time. We are learning which species are truly rare and which just appeared that way because only a few specialists were looking for them. Eventually we should be able to watch distribution maps of fungi shift in the same way we now watch weather patterns. A dispersed community of interested naturalists can monitor species of interest much more effectively than a small number of border officials and field scouts, something that is already happening with invasive plants. The energy of citizen scientists helps to carry knowledge about microfungi gained from the Petri dish back into nature to help find answers to that critical question: "What does it do?"

Apart from wandering in the wilds, there are other ways you can enjoy the benefits of the mycelial revolution. If you have a garden, growing native plants—which are ideally suited for conditions in your area—helps maintain the integrity of the ecosystem you call home. By cultivating these plants, you also preserve native microbial biodiversity. The mutualistic endophytes and mycorrhizae evolved along with these plants and should also be ideal for your local conditions. To optimize growth and keep weeds at bay, look for biofertilizers and biocontrol agents at garden centers or online. You can also buy little tins and bags of AM inoculum for flowers and shrubs in your garden. And for your lawn, look for premium grass seed that includes endophytes that protect the plants from hungry insects without having to use insecticides.

If you are able, contribute to the Trillion Tree Campaign.[18] This project aims to reduce atmospheric carbon by locking it away in the cellulose and lignin of a trillion newly planted trees. One of the founders is a mycorrhizal researcher. The success of this initiative will depend on endophytes and ectomycorrhizae, of course. Eventually, the project will also benefit the wood decay fungi, but preferably after a few centuries pass and we have more leeway with our carbon budget.

One Health: A Vision for a Sustainable World

Economic or political theory often portrays problems as a choice between two solutions, but our civilization is an interdependent network of collaborating and conflicting interests, many of them beyond human perception or control. The One Health movement promoted by the World Health Organization in partnership with many national bodies acknowledges that human health, the environment, and biodiversity are an interdependent triad.[19] In other words, robust communities of plants, animals, fungi, and bacteria contribute to healthy, productive, and resilient ecosystems. And to survive and prosper, humans depend on the health of these ecosystems—forests, farms, cities, and others—and the species that live in and modify them. One Health is built upon a network of treaties (conventions), action plans (protocols), and institutions that are responsible for addressing these overlapping priorities. For example, the World Health Organization, national and regional Centers for Disease Control, and the lesser-known World Organisation for Animal Health oversee aspects of human and animal welfare. These agencies actively monitor most of the fungal diseases that affect humans and they share information, especially with countries that lack expertise in medical or veterinary mycology. As data and knowledge grow, they will ensure that our microbiomes and mycobiomes are managed to maximum benefit.

The environment part of the triad is subject to many international initiatives, the best-known being the United Nations Framework Convention on Climate Change (with the Kyoto Protocol and subsequent accords as its action plan). A complex network of related national, provincial or state, and civic laws and regulations are often inconsistently applied in different jurisdictions. Still, together, these initiatives aim to stabilize global environmental systems to ensure a sustainable future, including improving our carbon balance and reducing pollution. Both are tightly bound with fungi. Part of the plan focuses on quarantine and preventing potentially invasive species (including fungi) from threatening entire ecosystems.

The Convention on Biological Diversity provides the backbone for protecting and conserving nature and educating citizens about Earth's biological heritage. The Convention on International Trade in Endangered Species of Wild Fauna and Flora (CITES) coordinates international protection of 669 animal and 334 plant species considered in danger of immediate extinction. Fungi generally need not apply, but two lichens, the rock gnome lichen (*Gymnoderma lineare*) and the Florida perforate cladonia (*Cladonia perforata*), are listed as endangered in the United States. Will social media ever try to crowdfund either of them back to health? The International Union for Conservation of Nature publishes the Red List of Threatened Species, with the International Society for Fungal Conservation contributing risk assessments for species considered at risk on the Global Fungal Red List. That list is now approaching 1,000 species, most of them mushrooms.[20]

Beyond lists of countries, institutions, agreements, and action plans, One Health is a philosophy that focuses collaborators from different economic sectors, and at every scale from local to global, on a shared vision of a sustainable future. It is also a multidisciplinary initiative encompassing public health, the environment, biodiversity, scientists, citizens, doctors, politicians, economists, government institutions, special interest groups, and entrepreneurs within this shared vision. One Health is a kind of mutualistic symbiosis at a policy-making level. As much as possible, we all need to work towards mutualism.

Reconsidering our attitude towards fungi is an important part of modifying our actions. I hope more people will become curious about our microscopic neighbors—or at least less suspicious or fearful of them. Fungi are among our closest relatives, and we are already deeply embedded with them. We should work with them a lot more than we do now. The future is fungal. It is also bacterial, algal, protistan, viral, buggy, wormy—full of all sorts of creatures, the big and beautiful, the small and ugly. Most of the life-forms around us were here long before we arrived and will remain long after we are gone. Let's learn what we can from them and hope for a long, rich journey together.

ACKNOWLEDGMENTS

THIS BOOK began with an email out of the blue from Jennifer Croll, Editorial Director at Greystone Books. I am grateful to Jennifer and her colleagues for their encouragement as this project grew, and to Ashlynne Merrifield (Radcliffe Cardiology) for her comments on the draft proposal. I learned a lot about the editorial processes behind "real books" from Linda Pruessen, who nurtured my early chapters to respectability. Lucy Kenward was as much a collaborator as an editor in bringing the book to its final form, and I will be forever grateful for her patience, insight, and tolerance for my caffeine-induced rants and sometimes adolescent sense of humor. Jessica Sullivan designed a wonderful cover and molded my cartoons into stylish vignettes with her design acumen. The insightful copy edit by Dawn Loewen gave the book a final polish.

Early in the process of planning this project, I decided not to name living mycologists, partly because I did not want this to be a book about me and my human friends. Despite that, with their permission, some of them do appear as friendly ghosts in the background. I am grateful to Charlene Hogan,

Dave Malloch, David Miller, Linda Payne, and Richard Summerbell, who commented on earlier drafts of the book. Matt Nelsen and Joey Tanney provided valuable input on specific chapters. Other colleagues clarified specific details or provided insight, including Jan Dijksterhuis, Mark Goettel, Gareth Griffiths, Sarah Hambleton, André Lévesque, Brent McCallum, Henrik Nilsson, Scott Redhead, and Franck Stefani. I benefited from the excellent library resources of Carleton University, the online Biodiversity Heritage Library, and consultations with the staff at the Uppsala University Archives and the International Churchill Society. Any errors that remain are, of course, my own.

Love and appreciation to Charlene for all her support during our year of quarantine, which I escaped by working on this book. Scratches under the chin and hugs for Rebus, who kept me moving over those months and who is now sorely missed by us both. There are several sets of three sisters in my life, including my own, who can interpret my dedication as they wish.

The writing of this book was supported by a grant from the Alfred P. Sloan Foundation's Public Understanding of Science, Technology and Economics program. My thanks to Paula Olsiewski, Doron Weber, and their staff for their encouragement.

APPENDIX
Fungal Classification

HERE YOU WILL FIND the classification (taxonomy) of the fungi mentioned in this book, at the ranks of kingdom, phylum (sometimes called division in mycology, with names typically ending with -mycota), class (-mycetes), and order (-ales).

I truly would rather protect you from the nerdish topic of fungal classification. Even though I've spent most of my career working with fungal taxonomy, it still often feels like a violent contact sport rather than a science. Most other biologists find ways to discreetly leave the room when taxonomists get started.

One fungal species sometimes has several scientific names (binomials). Fungal taxonomy has a particularly bad reputation for frequent changes, and to make matters worse, until 2011 we often used different names for sexual and asexual states of the same species. A few examples where competing sexual and asexual names are still widely used on the internet are included below. In addition, if a species is reclassified in a different genus, the epithets get moved from one genus to another like a star athlete traded to another team. Often, modern study, especially when involving DNA, leads to the realization that what was once considered a single species is actually a group of several species that are difficult to distinguish. In general, I have followed the results of such taxonomic splitting in this book without further explanation, but some situations where one fungus has several aliases in common use are noted below. This naming merry-go-round happens in many biological disciplines but gets dizzying in mycology. It turns the technical scientific literature into an impenetrable thicket for laypeople who want to learn about fungi from the internet or old textbooks.

In the list that follows, the approximate number of cataloged species for each group is based on estimates from catalogueoflife.org (version 28, August 2021).

KINGDOM *STRAMENOPILES* – oomycetes, water moulds (~1,700 species); not fungi from a modern evolutionary perspective, but traditionally studied by mycologists

> *Phytophthora infestans* – potato blight
> *Pythium* species – root diseases

KINGDOM *MYCOTA* – the true fungi (~146,000 species)

PHYLUM *Microsporidia* (~1,300 species)

> *Nosema bombycis* – silkworm pébrine disease
> *Nosema* species – locust biocontrol, microsporidiosis in people with AIDS

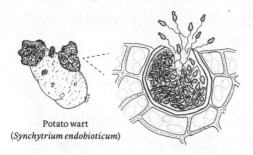

Chytridiomycota
A phylum of microscopic, often single-celled fungi producing zoospores and hypha-like structures called rhizoids, including pathogens of aquatic animals and plant roots.

Potato wart
(*Synchytrium endobioticum*)

PHYLUM *Chytridiomycota* – chytrids (~1,050 species)

> *Batrachochytrium dendrobatidis* (see figure on p. 201) –
> Bd, amphibian pathogen causing chytridiomycosis
> *Batrachochytrium salamandrivorans* – Bsal, salamander
> pathogen causing chytridiomycosis
> *Synchytrium endobioticum* (see figure above) – potato wart

PHYLUM *Neocallimastigomycota* – anaerobic fungi (35 species)

> *Neocallimastix frontalis* – cellulase-producing rumen anaerobe

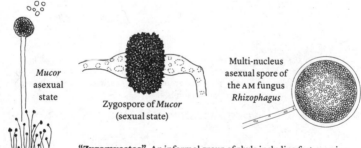

Mucor
asexual
state

Zygospore of *Mucor*
(sexual state)

Multi-nucleus
asexual spore of
the AM fungus
Rhizophagus

"Zygomycetes" An informal group of phyla including fast-growing mould-like fungi, endomycorrhizal species, and insect pathogens.

PHYLUM *Mucoromycota* (~300 species, previously in the now defunct phylum *Zygomycota*. Some AM fungi in class *Endogonomycetes* are also in this phylum.)

> *Mucor mucedo* – "hair of the cat" mould on cheese rind
> *Rhizomucor miehei* – fungal rennet
> *Rhizopus oligosporus* – tempeh mould
> *Rhizopus stolonifer* (see figure on p. 112) – compost mould

PHYLUM *Zoopagomycota* (~225 species, previously in the now defunct phylum *Zygomycota*)

> *Entomophthora muscae* – housefly fungus
> *Smittium morbosum* – mosquito pathogen

PHYLUM *Glomeromycota* – arbuscular mycorrhizal (AM) fungi (~335 species, previously in the now defunct phylum *Zygomycota*)

> *Funneliformis mosseae* – biofertilizer
> *Rhizophagus intraradices* – biofertilizer
> *Rhizophagus irregularis* – biofertilizer

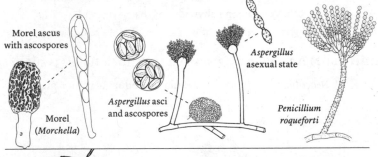

Morel ascus with ascospores

Morel (*Morchella*)

Aspergillus asci and ascospores

Aspergillus asexual state

Penicillium roqueforti

Truffle (*Tuber*)

Ascomycota A phylum of microfungi and macrofungi producing ascospores in asci, and many moulds, including many plant and animal pathogens, mycotoxin-producing species, and saprobes.

PHYLUM *Ascomycota* – ascomycetes or "ascos" (~93,000 species)

SUBPHYLUM *Taphrinomycotina* (~150 species)

> *Pneumocystis carinii* – Pneumocystis pneumonia in dogs
> *Pneumocystis jirovecii* – Pneumocystis pneumonia in people with AIDS

SUBPHYLUM *Saccharomycotina* (~1,200 species)

CLASS *Saccharomycetes* – "true" yeasts (~1,200 species)

> *Blastobotrys adeninivorans* – Pu-erh tea fermentation
> *Brettanomyces* species – Belgian beer fermentation

Candida albicans (see figure on p. 156) – candidiasis or thrush in humans, human mycobiont

Candida auris – candidiasis in humans

Candida parapsilosis – coffee fermentation

Candida tropicalis – candidiasis in humans

Cyberlindnera jadinii (= *Candida utilis*) – torula single-cell protein

Debaryomyces hansenii – cheese and meat fermentation, Crohn's disease

Geotrichum candidum – rinds of Camembert and Brie cheeses

Kloeckera species – chocolate fermentation

Kluyveromyces species – chocolate fermentation

Pichia nakasei – coffee fermentation

Saccharomyces boulardii – probiotic

Saccharomyces cerevisiae (see figure on p. 112) – brewer's and baker's yeast, human symbiont

Zygosaccharomyces rouxii – salty yeast

SUBPHYLUM *Pezizomycotina* (~91,500 species)

CLASS *Eurotiomycetes* – moulds, some lichens (~3,800 species)

ORDER *Chaetothyriales* (~700 species)

Exophiala species – black yeasts

ORDER *Eurotiales* (~1,350 species)

Aspergillus flavus – aflatoxin mould

Aspergillus fumigatus – human aspergillosis

Aspergillus glaucus (see figure on p. 1) (= *Eurotium herbariorum*, defunct name for sexual stage) – xerophilic food spoilage and house dust

Aspergillus niger – Pu-erh tea fermentation, production of citric acid and amylase

Aspergillus oryzae – kōji mould

Aspergillus terreus – itaconic acid fermentation, production of statins

Penicillium bilaiae – biofertilizer

Penicillium camemberti – Camembert and Brie cheese
Penicillium caseifulvum – Camembert and Brie cheese
Penicillium chrysogenum (= *P. notatum*) – penicillin fungus,
   indoor xerophile
Penicillium commune – dairy product spoilage
Penicillium digitatum – citrus rot
Penicillium expansum – blue mould of apples
Penicillium roqueforti (see figure on p. 224) – blue cheese,
   food and silage spoilage
Penicillium rubens – penicillin fungus
Thermomyces lanuginosus – thermophile, production of
   detergent enzymes

ORDER *Onygenales* (~425 species)

Blastomyces dermatitidis, B. gilchristii – human blastomycosis
   (Gilchrist's disease)
Coccidioides immitis – human coccidioidomycosis (valley fever)
Histoplasma capsulatum – human histoplasmosis
   (Caver's disease)
Trichophyton rubrum – athlete's foot, jock itch, ringworm

CLASS *Lecanoromycetes* – lichens (~8,000 species)

Cladonia perforata – Florida perforate cladonia (endangered)
Cladonia stellaris – star-tipped reindeer lichen
Gymnoderma lineare – rock gnome lichen (endangered)
Parmelia species – crottle, lichen dyes
Roccella tinctoria – litmus lichen
Umbilicaria species – rock tripe

CLASS *Leotiomycetes* – "inoperculate discomycetes," lots of
endophytes (~10,000 species)

Botrytis cinerea – noble rot of grapes
Hymenoscyphus fraxineus – ash dieback
Lophodermium species (see figure on p. 57) – conifer needle
   endophytes

Phialocephala species – conifer needle endophytes
Uncinula necator – powdery mildew of grapes

CLASS *Pezizomycetes* – "operculate discomycetes," "true truffles" (~2,800 species)

Chromelosporium fulvum – peat mould
Morchella species – morels
Tuber magnatum – white or Piedmont truffle
Tuber melanosporum – black or Périgord truffle

CLASS *Dothideomycetes* – "bitunicate ascomycetes," "loculoascomycetes" (~31,000 species)

Alternaria alternata – indoor dark mould
Aureobasidium pullulans – black yeast
Cladosporium cladosporioides – indoor dark mould, radioactivity-loving mould

CLASS *Sordariomycetes* – "pyrenomycetes," flask fungi (~23,000 species)

ORDER *Diaporthales* (~3,600 species)

Cryphonectria parasitica – Cp, chestnut blight

ORDER *Glomerellales* (~575 species)

Colletotrichum acutatum – weed biocontrol
Colletotrichum gloeosporioides – weed biocontrol

ORDER *Hypocreales* (~5,300 species)

Beauveria bassiana – white muscardine
Claviceps purpurea – ergot - - - - - - - - - -
Cordyceps species – zombie fungi
Epichloë species – grass endophytes
Escovopsis species – leaf-cutting ant nest parasite
Fusarium species – plant diseases
Fusarium graminearum – grain diseases, produces vomitoxin
(the old name of its sexual stage, *Gibberella*, is still sometimes used as a common name for some diseases)

Fusarium oxysporum – weed biocontrol

Fusarium venenatum – mycoprotein production

Fusarium verticillioides – produces fumonisin

Geosmithia morbida – thousand cankers disease of
   black walnut

Metarhizium acridum – green muscardine, insect
   biological control

Ophiocordyceps unilateralis – ant zombie fungus

Stachybotrys chartarum (see figure on p. 139) (= *S. atra* or
   S. atrus) – stachy, toxic black mould

Tolypocladium inflatum (see figure on p. 177) – cyclosporin A
   production

Trichoderma reesei – production of cellulases and other
   enzymes

Trichoderma virens – biofertilizer

Trichoderma species – industrial enzyme production,
   biocontrol

Trichothecium roseum – pink mould on cheese, mycotoxins

ORDER *Magnaporthales* (~280 species)

Gaeumannomyces tritici – take-all (disease of wheat)

ORDER *Ophiostomatales* (~400 species)

Grosmannia clavigera (= *Ophiostoma clavigerum*) – associate of
   mountain pine beetle

Leptographium longiclavatum – associate of mountain pine
   beetle

Ophiostoma novo-ulmi – Dutch elm disease (2nd wave)

Ophiostoma ulmi – Dutch elm disease (1st wave)

ORDER *Sordariales* (~1,400 species)

Chaetomium globosum – cellulase production, dark indoor
   mould

Neurospora crassa – red bread mould

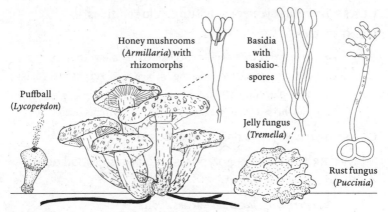

Puffball
(*Lycoperdon*)

Honey mushrooms
(*Armillaria*) with
rhizomorphs

Basidia
with
basidio-
spores

Jelly fungus
(*Tremella*)

Rust fungus
(*Puccinia*)

Basidiomycota A phylum of mostly macroscopic saprobic and
ectomycorrhizal fungi and microscopic plant pathogens, producing sexual
basidiospores from cells called basidia.

PHYLUM *Basidiomycota* – basidiomycetes or "basidios"
(~50,000 species)

CLASS *Wallemiomycetes* (~10 species)

> *Wallemia sebi* – domestic xerophile ---------

CLASS *Microbotryomycetes* (~325 species)

> *Rhodotorula taiwanensis* – radioactivity-
> loving yeast

CLASS *Pucciniomycetes* – rusts (~8,400 species)

> *Cronartium ribicola* – white pine blister rust
> *Hemileia vastatrix* – coffee rust
> *Puccinia graminis* (see figure above and on p. 87) – wheat rust
> (Ug99 race)

CLASS *Malasseziomycetes* (~25 species)

> *Malassezia* species (see figure on p. 156) – dandruff, mammal
> mycobionts

CLASS *Ustilaginomycetes* – smuts (~1,425 species)

> *Ustilago maydis* (see figure on p. 34) – corn smut, *huitlacoche*

CLASS *Tremellomycetes* – jelly fungi and relatives (~620 species)

> *Cryptococcus gattii* – cryptococcosis in people with AIDS
> *Cryptococcus neoformans* (see figure on p. 156) – cryptococcosis in people with AIDS
> *Tremella mesenterica* (see figure on p. 229) – witch's butter

CLASS *Agaricomycetes* (~38,300 species)

ORDER *Polyporales* – polypores, shelf fungi, wood decay fungi (~3,800 species)

> *Ganoderma applanatum* – artist's conk
> *Ganoderma lucidum* – reishi mushroom

ORDER *Cantherellales* – chanterelles (~850 species)

> *Cantharellus cibarius* – golden chanterelle (eastern NA), ectomycorrhizae
> *Cantharellus enelensis* – golden chanterelle (eastern NA), ectomycorrhizae
> *Cantharellus formosus* – golden chanterelle (western NA), ectomycorrhizae
> *Cantharellus phasmatis* – golden chanterelle (eastern NA), ectomycorrhizae

ORDER *Boletales* – boletes (~2,300 species)

> *Boletus edulis* – porcini, cep, Steinpilz
> *Serpula lacrymans* – dry rot
> *Spongiforma squarepantsii* – the SpongeBob mushroom

ORDER *Agaricales* – mushrooms, coral fungi, puffballs, etc. (~24,500 species)

> *Agaricus bisporus* – commercial mushroom
> *Amanita bisporigera* – destroying angel
> *Amanita phalloides* – death cap
> *Amanita virosa* – destroying angel

Armillaria gallica – bulbous honey mushroom, the 1st
humongous fungus

Armillaria mellea – honey mushroom

Armillaria ostoyae – honey mushroom, the 2nd humongous
fungus

Calvatia gigantea – giant puffball

Cortinarius orellanus – contains the kidney toxin orellanin

Leucocoprinus birnbaumii – flowerpot parasol

Leucocoprinus gongylophorus – leaf-cutting ant symbiont

Pleurotus ostreatus (see figure on p. 177) and related species –
oyster mushrooms

Psilocybe species – magic mushrooms

Schizophyllum commune – split gill mushroom (with 23,000
genders)

Tricholoma magnivelare – pine mushroom or matsutake

Tricholoma matsutake (= *T. nauseosum*) – pine mushroom or
matsutake

ORDER *Russulales* – (~3,270 species)

Lactarius species – milk caps, ectomycorrhizae

Russula species – brittle gills, ectomycorrhizae

NOTES

A Note About Names

1. Whether there is a soft or hard *g* in "fungi" depends on whom you ask or what dictionary you check. Current editions of the *Oxford English Dictionary* and *Merriam-Webster's Collegiate Dictionary* accept both, which is good enough for me, but the *Cambridge Dictionary* is hard-nosed about insisting on a hard *g*.
2. List of mushroom common names for the U.K.: Holden (2003); for North America, see the websites for the North American Mycological Association (namyco.org) and iNaturalist (inaturalist.org).
3. "Kingdom" is the term traditionally used for the largest groupings of life (animals, plants, fungi, etc.). It is difficult to compare across kingdoms, but for the purposes of comparison with the animal kingdom, chordates (animals with backbones) and arthropods (insects and other creatures with exoskeletons) are different phyla; within chordates, mammals and birds are different classes; and within mammals, primates and rodents are different orders. This relative or hierarchical classification does not correlate with genetic divergence. Fungi may encompass ten times more genetic diversity than animals.
4. For amusing Latin names, see the website Curiosities of Biological Nomenclature, curioustaxonomy.net/puns/puns.html.

Introduction

1. Amato (2001) and Holmes (2003) consider the physical and biological composition of dust, and our cultural and psychological attitudes towards it. Griffin et al. (2001) gives an overview of the public and

ecosystem health consequences of the transcontinental spread of dust.

2. Microbial diversity revealed by fluorescent staining: Suttle (2013).

3. Rock tripe and the Franklin expeditions: Smith (1877).

4. *Mukēs* is the Greek word for "fungus," so mycology is the study of fungi. The British alternative "fungology" and "fungologist" never caught on, thankfully. The classic form gives us more scope for puns, like "Mycology is better than yours."

5. A beautiful photographic overview of macrofungi: Petersen (2013).

6. Mushroom murder mysteries: Wasson (1972).

7. The concept of mycophobia (as fungiphobia) is attributed to Hay (1887). The word "mycophile" has been used as a noun for amateur mushroom lovers in France since at least the 1880s (OED online, 2021). The terms were set into opposition as cultural characters by Wasson & Wasson (1957), which subsequently stimulated abundant ethnographic research.

8. For a scolding of the scientific prejudice against anthropomorphism, see de Waal (2019).

9. The nuances behind calculating the number of fungal species are discussed by Blackwell (2011).

1 | Life in the Colonies

1. Origins of life: Beerling (2019), Margulis & Sagan (1986). History of life and plate tectonics: David Christian's (2011) TED talk gives an engaging introduction to the history of the Earth, and Prosanta Chakrabarty's (2018) TED talk does the same for the history of life.

2. Last common ancestors: Among traditionally recognized kingdoms, fungi are much closer relatives to animals than plants. The number and composition of kingdoms, until recently considered the broadest category of classification of life, is revised frequently as more genome data for rare microbial groups becomes available. A few single-celled groups, previously considered protists, are intermediate between animals and fungi—for example, the choanoflagellates, a small group of about 125 microscopic species that live in water. See Burki (2014), Keeling & Burki (2019).

3. Fungal taxonomy: The details of morphology and life cycles for the major fungal groups are well known and verifiable from almost any general mycology textbook written for undergraduate university

students. The books I consulted most often during my career were Alexopoulos et al. (1996), Burnett (1976), Kendrick (2017), and Webster & Weber (2007). For a beautifully illustrated and non-academic overview of fungal morphology, it is impossible to beat the book by the creative Danish photographer and mushroom enthusiast Jens Petersen (2013). Keep in mind that while older textbooks have their charm, those published in the previous century will miss out on the revolution initiated by DNA-based classification.

4. Mushroom identification: There are many excellent mushroom manuals. Any book you choose must have a good introduction to the methods used to accurately identify mushrooms. Look for a book published specifically for your country or region with color photographs and including at least 150 species. If you can find a local or national mushroom club, the members will recommend their favorites. For North American clubs, check the listings on the North American Mycological Association website (namyco.org/clubs.php). For Europe, check the European Mycological Association website (euromould.org/resources/links/socs.html).

5. Mushroom toxicity: Most mushroom manuals include detailed information on poisonous mushrooms. Mycological and medical aspects: Lincoff & Mitchel (1977). Chemistry of toxins: Yin et al. (2019).

6. Next-gen is now being replaced with new techniques called third-generation DNA sequencing, and next-gen is then sometimes referred to as second-generation.

7. If Google Earth didn't stop: Google "Powers of Ten" for the famous 1977 video by Charles and Ray Eames, inspired by Boeke (1957), which zooms back and forth from the subatomic scale to the entire known universe. The animation by Lariontsev about *Aspergillus* diverts such a zoom into a fungal version of reality (youtu.be/8xrB9boqDjg).

8. Life as a fungus: There is a long tradition of imagining life through the eyes of other animals—for example, Nagel's (1974) essay about what it is like to be a bat. Hoyt (1996) attempts such imaginings for ants, Sibley (2020) for birds, and Powers (2018), in his novel *The Overstory*, for trees. The fungal point of view is imagined in Piers Anthony's (1968) science fiction novel *Omnivore* and Brie Larson's narration of Louie Schwartzberg's (2019) film *Fantastic Fungi*.

9. Hyphae in soil: Ekblad et al. (2013).
10. Clones and self-recognition: Hall et al. (2010).
11. Spores: Mycologists are obsessive about spores, and most books about fungi will have hundreds of spore illustrations. Mycological texts tend to be crammed full of drawings and photographs of spores—one of my chemistry profs called it "spore-nography."
12. Speed of growth: *Neurospora crassa* grows 3 or 4 inches per day under ideal conditions (Ryan et al. 1943). Other fungi grow at much slower rates (Moore et al. 2020).
13. Dynamics of mycelial growth: Hale & Eaton (1985), Rayner (1997).
14. Genetics of fungal mating: Fraser & Heitman (2003).
15. The 23,000 genders (of *Schizophyllum commune*): Rokas (2018).
16. Fungal chemistry, pheromones, and other metabolites: Bills & Gloer (2016).

2 | Life on the Commons

1. Overview of symbiosis: Margulis (1998). Anton de Bary: See Ward's (1888) biography.
2. Darwin's blind spot: Hammerstein (2003), Ryan (2002).
3. Lichens in Sudbury: Beckett (1995). Lichens as pollution monitors: Hawksworth & Rose (1976).
4. There are several books for general readers about lichen biology. I used Richardson (1975) as a starting point. Although not as well served as the mushroom community, lichenologists enjoy some awe-inspiring identification manuals based on visual characteristics, my favorite being Brodo et al. (2001). The website associated with that book includes a comprehensive compilation of the folk use of lichens through the millennia (sharnoffphotos.com/lichen_info/usetype.html). A recent view of lichen diversity and evolution is provided by Nelsen et al. (2020).
5. Lichenometry and the age of Arctic lichens: Armstrong (2015). An age of 8,600 years is often quoted online, although I could not find published data to support this.
6. Weathering of rocks: Chen et al. (2000).
7. Lichen dyes: Casselman (2001).
8. Lichen secondary metabolites: Shrestha & St. Clair (2013).
9. Leaf-cutting ants: Hoyt (1996) deals with these ants often from an anthropomorphic point of view; Hölldobler & Wilson (2010) is an academic but very approachable text.

10. The fungus cultivated by leaf-cutter ants is a close relative of the flowerpot parasol, *Leucocoprinus birnbaumii*, a slender, scaly-capped yellow mushroom that sometimes pops out beside house plants in human nesting chambers. Perhaps it is waiting for us to domesticate it and offer it a better life.

11. Search YouTube for the video "The Ant City" to see the outcome of the filling of a leaf-cutter colony with concrete in Brazil (Patrick 2013).

12. Genomes and the ant-fungus symbiosis: Nygaard et al. (2016).

13. Trichomycetes: Lichtwardt (2012).

14. Pasteur and *Nosema*: Borst (2011).

15. American origin of grape powdery mildew: Brewer & Milgroom (2010). Bordeaux mixture: Large (1940).

16. Corn smut as a plant pathogen: Pataky & Snetselaar (2006). *Huitlacoche*: Lipka (2009).

17. Invasive species: Kolbert (2014) is a stirring call to arms. For a broader look, see Anthony (2017).

18. The Irish potato famine: Some of the material in this section appeared, in a different form, in Seifert (2013). See also Large (1940).

19. Biocontrol of weeds in Australia: Palmer et al. (2010).

20. Mitochondria and chloroplasts as symbionts: Margulis (1998), Martin & Mentel (2010).

3 | Forests

1. Endophytes: Pirttilä & Frank (2011).

2. *Lophodermium* as an endophyte: Tanney et al. (2018).

3. History of ectomycorrhizae: Trappe (2005).

4. Plants lacking mycorrhizae: Brundrett & Tedersoo (2018).

5. Review of ectomycorrhizal biology: Smith & Reid (2008).

6. Influence of mycorrhizae on plant diversity: Brundrett & Tedersoo (2018).

7. "Wood Wide Web" was first coined by British ecologist David Moore in an editorial accompanying the article by Simard et al. (1997) that demonstrated communication among trees in a natural forest. See also Simard (2021), Wohlleben (2016), various chapters in Sheldrake (2020).

8. Swedish work on tracing nutrient flow between tree roots and fungal hyphae: Lindeberg (1989).

9. Mother trees and mycorrhizae: Simard (2018, 2021).

10. The combined matsutake and chanterelle harvest sometimes exceeds the commercial value of timber from the same forests (Alexander et al. 2002).

11. Economic and cultural aspects of harvesting pine mushrooms: Tsing (2015). The fragrance "recalls times past": Tsing (2015, 48).

12. Truffles: One hundred metabolites: Mustafa et al. (2020); prices, McCutchen (2017); international marketing and fraud in the truffle trade, Jacobs (2019).

13. Humongous fungus: Anderson et al. (2018), Casselman (2007), Zhang (2017). *Armillaria mellea* was considered a variable species for a long time but is now divided into at least a dozen distinct species, two of which are *A. gallica* and *A. ostoyae*. They have similar appearances and ecologies, but they lead independent lives and do not mate with each other.

14. The challenges faced by Dutch female scientists working on Dutch elm disease are reviewed by Holmes & Heybroek (1990).

15. Dutch elm disease: D'Arcy (2000). The two waves were at first both blamed on *Ophiostoma ulmi*, but when genetic tests showed that the pathogens from the first two waves could not mate with each other, they were acknowledged as separate species.

16. Chestnut blight: Anagnostakis (1987).

17. CHV1: *Hypo* means "reduced," and fungal strains infected with the virus don't spread as easily and are less virulent. Incidentally, SARS-CoV-2 (the virus that causes COVID-19) is also an RNA virus.

18. Fungal viruses: Ghabrial et al. (2015).

19. White pine blister rust: Geils et al. (2010). Ash dieback: CABI Invasive Species Compendium (n.d.). Thousand cankers disease: thousandcankers.com.

20. Plant biomass: Bar-On et al. (2018).

21. Wood decay: Rayner & Boddy (1988).

22. In tropical countries, some mycologists train monkeys to climb trees to collect specimens.

23. Endophytes as biological control agents: Tanney et al. (2018).

4 | Farming

1. The "Seventh-Oldest Profession" in the chapter title comes from a sober evaluation of what *actually* are the world's oldest professions; see Oldest.org (n.d.).

2. History of agriculture: Diamond (1997) and Harari (2014) provide contrarian views of the agricultural revolution, and some of their ideas about domestication are reflected in this book. Large (1940) is an exhaustive but fascinating historical review of the involvement of fungi in agriculture.
3. Ammonium is made up of nitrogen and hydrogen atoms. Nitrogen fixation: Wagner (2011).
4. For spectacular photographs of arbuscular mycorrhizae (AM) fungi made using confocal microscopy, see Kokkoris et al. (2020) or check out Vasilis Kokkoris's website (vasilis-kokkoris.com).
5. History of the discovery of AM fungi and early research: Koide & Mosse (2004).
6. Interaction of AM fungi with wheat: Fiorilli et al. (2018). AM fungi and plant responses to stress: Begum et al. (2019). Common mycorrhizal networks and AM plants: Wipf et al. (2019).
7. Increasing phosphorus uptake using *Penicillium*: Leggett et al. (2007).
8. Grass endophytes: Bacon (2018), Clay (1990), Schardl & Phillips (1997).
9. Harari (2014) questions whether humanity domesticated wheat, or it domesticated us. Rust (see next section of the chapter) is such an intimate, targeted disease, we can add a dimension and ponder its contribution to this process.
10. Efforts to eliminate barberry: Mann (2018). "Execute this criminal bush": Mann (2018, 114).
11. Review of rust: Schumann & Leonard (2011).
12. *Explorer II* flight: Kennedy (1956).
13. Wheat breeding: McCallum & DePauw (2008). When the resistance genes of the crop plant match the pathogenicity genes of the rust fungus, there is no disease. This is called one-for-one gene resistance. It is as if the plant has a gene that locks the rust out. It is analogous to the lock-and-key relationship between variants of human viral diseases and antibodies in vaccines.
14. Green Revolution: Mann (2018) vividly reconstructs the Green Revolution and the conflicts between environmentalists and industrialized agriculture. The stories about Norman Borlaug and rust in India are given in full detail in Mann's book, which also includes interesting information on the potato famine, rusts, ergot, and plant breeding against fungal diseases in general.

15. Ug99: Pretorius et al. (2000).
16. Charles Tulasne was a brilliant artist, and his vivid drawings gave him a reputation as the "Audubon of fungi."
17. Review of ergot: Schumann & Uppala (2017).
18. History of ergot toxicity: Matossian (1982, 1991).
19. Discovery of LSD, microdosing of psychedelics: Pollan (2018). To learn more about stoned apes, search out the film *Fantastic Fungi* (Schwartzberg 2019) or the accompanying book (Stamets 2019).
20. Food contamination statistics: See the FAO's website Technical Platform on the Measurement and Reduction of Food Loss and Waste, fao.org/food-loss-and-food-waste. Mycotoxins: Pitt et al. (2012).
21. Aflatoxin: Kumar et al. (2017), Schrenk et al. (2020).
22. Vomitoxin: Sobrova et al. (2010).
23. Fumonisins: Marasas (1995). See also World Health Organization (2018).
24. There is no shortage of information about zombie fungi on the internet. The YouTube video "Cordyceps: Attack of the killer fungi" (BBC Studios 2008) illustrates the phenomenon in graphic time lapse. Sheldrake (2020, chapter 4) and Yong (2017) will lead you to the research papers of David Hughes at Penn State, among others.
25. Agostino Bassi: Porter (1973).
26. Biocontrol using *Beauveria*: García-Estrada et al. (2016).
27. Biocontrol of grasshoppers and locusts: Lomer et al. *(*2001).
28. Use of fungi for biocontrol: Butt et al. (2001).
29. Nine major plant diseases: Savage (n.d.).

5 | Fermentation

1. Two recent books stretch the stories of yeast and alcohol to feature length. Money (2018) gives a general overview of the science, history, and social dimensions of humanity's relationship with yeasts. Rogers (2015) fulfills a similar function for alcoholic beverages. Both books comment on the coincidental emergence of agriculture and brewing technology.
2. My calculations of the number of yeast cells and the relative numbers of stars and humans are based on various sources in the scientific literature. If a yeast cell lives four days, there are 91.5 generations per year. Estimates of the number of stars in the Milky Way galaxy top out at about 400 billion.

3. Asian origin of brewer's yeast: Duan et al. (2018).
4. Yeast hybridization and genomics: Peter et al. (2018).
5. Cork taint in wine: Álvarez-Rodríguez et al. (2002).
6. History of unleavened bread: Arranz-Otaegui et al. (2018).
7. Bread yeasts: Carbonetto et al. (2018), Money (2018).
8. Microbiology of sourdough: Carbonetto et al. (2018), Dunn (2018). The sourdough library is described by Ewbank (2018).
9. Artisanal cheeses: See the website Microbialfoods.org. *Geotrichum* in cheese: Boutrou & Guéguen (2005). "Yeast-like" is a term used for fungi that make a budding yeast form and some hyphae, or cells similar to hyphae.
10. Exports of blue cheese: NationMaster (2019).
11. Mycotoxins in cheeses: Nielsen et al. (2006), Scott & Kanhere (1979).
12. Some products marketed as soy sauce are not fermented. If the ingredients include hydrolyzed protein and don't mention a brewing starter or culture, your sauce is an imposter produced by a nonfungal process.
13. Some mycologists consider *Aspergillus oryzae* and *A. flavus* the same species, and you sometimes see the latter name as an ingredient on soy sauce labels. Whatever the taxonomic confusion, soy sauce is aflatoxin-free.
14. Genomics of kōji: Gibbons et al. (2012).
15. History of kōji and domestication of *Aspergillus*: Shurtleff & Aoyagi (2012).
16. Yeast fermentation and chocolate: Ludlow et al. (2016).
17. Pu-erh tea: Abe et al. (2008).
18. Coffee fermentation: de Oliveira Junqueira et al. (2019).
19. Coffee rust: Large (1940), McKenna (2020).
20. Food spoilage: Dijksterhuis & Samson (2007).
21. Food waste data: See the FAO's Food Loss and Waste Database, fao.org/food-loss-and-food-waste/flw-data.
22. Xerophiles: Pettersson & Leong (2011). The xerophilic *Aspergillus* species are often called *Eurotium*.
23. Food preservation: Eschliman & Ettlinger (2015) give a colorful overview of some of the chemicals used in food. Not for the chemophobic.
24. Compost: Wright et al. (2016).
25. Home mushroom cultivation: Stamets & Chilton (1983).

26. Although tempeh is considered a traditional Indonesian food, it's only been made for about two hundred years. Shurtleff & Aoyagi (1979) offer a comprehensive historical and nutritional review.

6 | The Secret House

1. There is abundant misinformation about moulds in houses on the Web. There are many technical books on indoor moulds, of highly variable quality; Flannigan et al. (2003) is authoritative. Bodanis (1986; the origin of this chapter's title) and Dunn (2018) both give entertaining reviews of various aspects of the science and biology of the built environment from a nontechnical (mostly nonfungal) perspective. Schilthuizen (2018) presents an overview of the adaptations of biodiversity to cities, focusing on birds, insects, and plants.
2. Fungal odors: Horner & Miller (2003). Some building inspectors use dogs trained to sniff out the telltale fungal volatiles. One wonders about the ethics of this; I wouldn't train *my* dog for this job.
3. Sap stain: See chapter 14 in Zabel & Morrell (2012).
4. Dry rot: VanderGoot (2017).
5. Wood preservatives: Chapter 19 in Zabel & Morrell (2012).
6. Moulds in dishwashers: Zupančič et al. (2016).
7. Urbanization of *Aureobasidium*: Schilthuizen (2018).
8. DNA surveys of house dust: This paragraph describes an unexplored aspect of the next-gen data from the same samples studied by Amend et al. (2010). Dunn (2018) summarizes later DNA surveys of all kinds of organisms occurring in houses, not to mention the International Space Station, but focuses more on bacteria than fungi.
9. *Wallemia*: Desroches et al. (2014).
10. The official chemical name for this noxious polysaccharide is $(1{\rightarrow}3)$-β-D-glucan. See Maheswaran et al. (2014).
11. Domestic moulds, asthma, and allergies: Rosenblum Lichtenstein et al. (2015), Tischer et al. (2011). Dust mites, fungi, and allergies: Miller (2019), Van Asselt (1999).
12. *Stachybotrys* in buildings: Miller et al. (2003).
13. The efficiency of HEPA filters is tested and products are given a minimum efficiency reporting value (MERV) between one and sixteen, the higher the better. See EPA (2021).
14. Hygiene and remediation: For reliable advice, the *Mold Remediation in Schools and Commercial Buildings* guide (EPA 2008) is an

excellent starting point. I have also often consulted the Canadian Construction Association's (2018) mould guidelines and appreciated their precise, pragmatic advice for remediating mould situations.

7 | Holobiont

1. Human Genome Project: Chial (2008).
2. Human Microbiome Project: Gevers et al. (2012). Microbiomes: Yong (2016) is an entertaining walk through microbiomics, mostly from a bacterial perspective, and gives a broad introduction to the emerging paradigm of holobionts. Animals with less extensive microbiomes: Cepelewicz & Quanta Magazine (2020).
3. The number of human cells is based on estimates by Bianconi et al. (2013). Relative number of human and bacterial cells in the human microbiome: Sender et al. (2016).
4. Fungi and the extinction of dinosaurs: Casadevall (2012). Mycobiomes: Cui et al. (2013), Enaud et al. (2018), Seed (2015).
5. Shedding of skin cells: AAAS (2009).
6. Biology of *Malassezia*: Saunders et al. (2012).
7. Skin infections: Gräser et al. (2018).
8. *Saccharomyces* as a probiotic: Czerucka et al. (2007).
9. *Candida* as a symbiont: Hall & Noverr (2017).
10. Histoplasmosis: Chapter 12 in Cordeiro (2019), Kauffman (2007). Blastomycosis: Schwartz & Kauffman (2020). Coccidioidomycosis: Kirkland & Fierer (2018).
11. The Latin name *fumigatus* means "smoky."
12. Allergic bronchopulmonary aspergillosis: See the Aspergillus & Aspergillosis website, aspergillus.org.uk.
13. Fungal diseases and AIDS: Centers for Disease Control and Prevention (2020b), Limper et al. (2017), UNAIDS (2021). Because of different statistical and reporting methods used at different times and in different places, there is often a lack of consistency in these estimates.
14. *Pneumocystis*: Sokulska et al. (2015). *Pneumocystis* names: Stringer et al. (2002).
15. *Cryptococcus*: May et al. (2016).
16. Fungal diseases of humans: The estimates for the number of infected humans are from Bongomin et al. (2017). The website

of the International Society for Human and Animal Mycology (isham.org) is the starting point for reliable information and links regarding fungal infections of humans. Cordeiro's (2019) book summarizes the diagnoses of fungal infections.
17. Amphotericin B: Laniado-Laborín & Cabrales-Vargas (2009).
18. Antibiotics and Crohn's disease: Jain et al. (2021). Dysbiosis: Bäckhed et al. (2012).

8 | Mycotechnology

1. "Mycelial Revolution" in the title for part 3 echoes the phrase "mycelium revolution," coined by myco-entrepreneur Eben Bayer (2019). The word "mycotechnology" was introduced by Bennett (1998).
2. Biotechnology and *Aspergillus niger*: Cairns et al. (2018).
3. The scientific name of the penicillin fungus has changed several times, from *P. rubrum* to *P. notatum*, *P. chrysogenum*, and finally *P. rubens* (Houbraken et al. 2011).
4. Fleming (1929) is the famous paper describing penicillin.
5. The whole penicillin story is brilliantly told by Lax (2004). A closet-like museum now sits where Fleming made his discovery in St. Mary's Hospital, London. It was returned to its former state from photographs, then allowed to become as dusty as it apparently was in Fleming's time. Fleming's grave is in the lower level of St. Paul's Cathedral, if you care to pay your respects—I have.
6. Streptomycin: Pringle (2013).
7. Cyclosporin A: Heusler & Pletscher (2001).
8. Antibiotic resistance in medicine: Centers for Disease Control and Prevention (2020a), World Health Organization (n.d.).
9. Rising resistance to azole drugs: Meis et al. (2016). Use in agriculture: Verweij et al. (2009).
10. Drugs made by genetically modified yeast: Nielsen (2013).
11. Development and marketing of Quorn: Wiebe (2004). Secondary metabolites of *Fusarium venenatum*: Miller & MacKenzie (2000).
12. Biotechnology of lignin: Irmer (2017).
13. Bioenergy and biofuels: Lange (2017), Salehi Jouzani et al. (2020).
14. Biotechnology using anaerobic gut fungi: Flad et al. (2020).
15. Mycoremediation: Dykes (2021). Oil spill cleanup: MatterofTrust.org (n.d.). Mushrooms and radioactive cesium:

Garaudée et al. (2002). *Cladosporium* on spacecraft: Shunk et al. (2020; the allergenic potential of the fungus seems not to have been considered). *Rhodotorula* and radioactivity: Tkavc et al. (2018).

16. "Generally recognized as safe" (GRAS) is a designation for certain microbes or chemicals that do not require additional evaluation and approval by the United States Food and Drug Administration before they are used in foods or drugs.

17. Plastic bioremediation: Sánchez et al. (2020).

18. Detoxification of coffee waste: Brand et al. (2020). Dye absorption using coffee wastes: Cheruiyot et al. (2019).

19. Two manifestos for a mycelial revolution are by Stamets (2005) and McCoy (2016). Paul Stamets's (2008) TED talk is a popular summary of his take on mycotechnology.

20. Most information about artisanal beer comes from press releases, and there is not much published, peer-reviewed literature on this topic. Akpan & Ehrichs (2017) discuss some brews based on insect yeasts; Metcalfe (2016) describes shipwreck beer. Other brewing yeasts: Holt et al. (2018).

21. Cheese mould domestication: Gibbons (2019). An example is a white-spored version of blue cheese, called Nuworld, developed in the United States.

22. New mycelial products: Cerimi et al. (2019).

23. A 2020 exhibit in London—*Mushrooms: The Art, Design and Future of Fungi*—was a stirring presentation of fungal art and the products of boutique fungal biotechnology (gaiaartfoundation.org/projects/mushrooms-the-art-design-and-future-of-fungi).

24. Mars myco-architecture: Malone (2018).

25. Fungal future: These actual and potential innovations come from the popular press and can be tracked down easily in text or video form; some are described at more length in other chapters and are referenced there.

9 | Thirty Thousand Feet

1. One explanation for crop circles is a fungal disease called "take-all" caused by the ascomycete *Gaeumannomyces tritici*.

2. Satellite monitoring of plant diseases: Oerke (2020).

3. Sustainable development goals: See the United Nations website The 17 Goals, sdgs.un.org/goals. Emerging threats: Fisher et al. (2016). Mass extinction: Kolbert (2014).

4. The term "Pangaeafication" seems to have originated with British botanist Mark Spencer (markspencerbotanist.com).

5. The labeling of humans as invasive is often debated, and which side you take may depend on how you interpret the United Nations Environment Programme's (2016, para. 6) definition that an invasive species is "destructive to human interests and natural systems."

6. Lists of regulated pests vary by country and change regularly. For a starting point, search the FAO's International Plant Protection Convention website (ippc.int).

7. Potato wart: Franc (2007).

8. Amphibian extinction: Charles (2021), Kolbert (2014, chapter 1). Frog chytrid origin: O'Hanlon et al. (2018).

9. Control of Bd using bacteria: Yong (2016).

10. Bsal: Grant et al. (2016).

11. Fortey (2008) provides an amusing backroom glimpse into the esoteric world of natural history collections.

12. Neither the CBD nor Nagoya was signed by the United States.

13. Ownership of biodiversity: Gepts (2004). Material collected before 1992 is excused from benefit sharing, or "grandfathered," under those treaties. But for older specimens, there are still frequent discussions about whether all biological cultural treasures should be repatriated.

14. Inoculation of fungi into Finnish forests: Abrego et al. (2016).

15. The phrase "Guardians of the Microbial Galaxy" is from Hariharan (2021).

16. Fungi and the bioeconomy: Meyer et al. (2020).

17. Online databases with occurrence information on fungi: iNaturalist (inaturalist.org, with many countries having their own subdomains); Global Biodiversity Information Facility (gbif.org); Mushroom Observer (similar to iNaturalist but only for fungi, mushroomobserver.org); Mycoportal (fungal specimens in North American biological collections, mycoportal.org).

18. Trillion Tree Campaign: trilliontreecampaign.org.

19. One Health: onehealthcommission.org.

20. Websites concerned with extinction and conservation: Convention on International Trade in Endangered Species (cites.org); IUCN Red List of Threatened Species (iucn.org/resources/conservation-tools/iucn-red-list-threatened-species); International Society for Fungal Conservation (fungal-conservation.org).

LITERATURE
CITED

Note: OA = open access (indicates scientific journal articles that are freely available online).

AAAS. 2009. *The Science Inside: Skin.* American Association for the Advancement of Science.

Abe M, Takaoka N, Idemoto Y, et al. 2008. Characteristic fungi observed in the fermentation process for Puer tea. *International Journal of Food Microbiology* 124: 199–203.

Abrego N, Oivanen P, Viner I, et al. 2016. Reintroduction of threatened fungal species via inoculation. *Biological Conservation* 203: 120–124.

Akpan N, Ehrichs M. 2017. The beers and the bees: Pollinators provide a different kind of brewer's yeast. *Scientific American*, June 26, 2017. scientificamerican.com/article/the-beers-and-the-bees-pollinators-provide-a-different-kind-of-brewer-rsquo-s-yeast.

Alexander SJ, Pilz D, Weber NS, et al. 2002. Mushrooms, trees, and money: Value estimates of commercial mushrooms and timber in the Pacific Northwest. *Environmental Management* 30: 129–141.

Alexopoulos CJ, Mims CW, Blackwell MM. 1996. *Introductory Mycology.* 4th ed. John Wiley & Sons.

Álvarez-Rodríguez ML, López-Ocaña L, López-Coronado JM, et al. 2002. Cork taint of wines: Role of the filamentous fungi isolated from cork in the formation of 2, 4, 6-trichloroanisole by O methylation of 2, 4, 6-trichlorophenol. *Applied and Environmental Microbiology* 68: 5860–5869.

Amato JA. 2001. *Dust: A History of the Small and the Invisible.* University of California Press.

Amend AS, Seifert KA, Samson R, et al. 2010. Indoor fungal composition is geographically patterned and more diverse in temperate zones than in the tropics. *Proceedings of the National Academy of Sciences* 107: 13748–13753.

Anagnostakis SL. 1987. Chestnut blight: The classical problem of an introduced pathogen. *Mycologia* 79: 23-37.

Anderson JB, Bruh JN, Kasimer D, et al. 2018. Clonal evolution and genome stability in a 2,500-year-old fungal individual. *Proceedings of the Royal Society B: Biological Sciences* 285: article 20182233 (OA).

Anthony L. 2017. *The Aliens Among Us: How Invasive Species Are Transforming the Planet—and Ourselves*. Yale University Press.

Anthony P. 1968. *Omnivore*. Ballantine Books.

Armstrong RA. 2015. Lichen growth and lichenometry. In Upreti DK, Divakar PK, Shukla V, et al. (eds), *Recent Advances in Lichenology: Modern Methods and Approaches in Biomonitoring and Bioprospection*, vol. 1, 213–227. Springer.

Arranz-Otaegui A, Carretero LG, Ramsey MN, et al. 2018. Archaeobotanical evidence reveals the origins of bread 14,400 years ago in northeastern Jordan. *Proceedings of the National Academy of Sciences* 115: 7925-7930 (OA).

Bäckhed F, Fraser CM, Ringel Y, et al. 2012. Defining a healthy human gut microbiome: Current concepts, future directions, and clinical applications. *Cell Host and Microbe* 12: 611–622 (OA).

Bacon CW. 2018. *Biotechnology of Endophytic Fungi of Grasses*. CRC Press.

Bar-On YM, Phillips R, Milo R. 2018. The biomass distribution on Earth. *Proceedings of the National Academy of Sciences* 115: 6506-6511 (OA).

Bayer E. 2019. The mycelium revolution is upon us. *Scientific American*, July 1, 2019. blogs.scientificamerican.com/observations/the-mycelium-revolution-is-upon-us.

BBC Studios. 2008. "Cordyceps: Attack of the killer fungi – Planet Earth Attenborough BBC wildlife." YouTube video, posted November 3, 2008, 3:03. (Clip from the documentary series *Planet Earth*, dir. Fothergill A, Linfield M, 2006). youtube.com/watch?v=XUKjBIBBAL8.

Beckett PJ. 1995. Lichens: Sensitive indicators of improving air quality. In Gunn JM (ed), *Restoration and Recovery of an Industrial Region: Progress in Restoring the Smelter-Damaged Landscape Near Sudbury, Canada*, 81–91. Springer.

Beerling D. 2019. *Making Eden: How Plants Transformed a Barren Planet.* Oxford University Press.

Begum N, Qin C, Ahanger M A, et al. 2019. Role of arbuscular mycorrhizal fungi in plant growth regulation: Implications in abiotic stress tolerance. *Frontiers in Plant Science* 10: article 1068 (OA).

Bennett J W. 1998. Mycotechnology: The role of fungi in biotechnology. *Journal of Biotechnology* 66: 101–107.

Bianconi E, Piovesan A, Facchin F, et al. 2013. An estimation of the number of cells in the human body. *Annals of Human Biology* 40: 463–471.

Bills G F, Gloer J B. 2016. Biologically active secondary metabolites from the fungi. *Microbiology Spectrum* 4(6).

Blackwell M. 2011. The fungi: 1, 2, 3 . . . 5.1 million species? *American Journal of Botany* 98: 426–438 (OA).

Bodanis D. 1986. *The Secret House: 24 Hours in the Strange and Unexpected World in Which We Spend Our Days and Nights.* Simon & Schuster.

Boeke K. 1957. *Cosmic View: The Universe in 40 Jumps.* John Day Co.

Bongomin F, Gago S, Oladele R O, Denning D W. 2017. Global and multi-national prevalence of fungal diseases—estimate precision. *Journal of Fungi* 3: article 57 (OA).

Borst P L. 2011. Silk, Pasteur and the honey bee: The story of Nosema disease. *American Bee Journal* 151: 773–777 (OA).

Boutrou R, Guéguen M. 2005. Interests in *Geotrichum candidum* for cheese technology. *International Journal of Food Microbiology* 102: 1–20.

Brand D, Pandey A, Roussos S, et al. 2020. Biological detoxification of coffee husk by filamentous fungi using a solid state fermentation ·system. *Enzyme and Microbial Technology* 27: 127–133.

Brewer M T, Milgroom M G. 2010. Phylogeography and population structure of the grape powdery mildew fungus, *Erysiphe necator,* from diverse *Vitis* species. *BMC Evolutionary Biology* 10: article 268.

Brodo I M, Sharnoff S D, Sharnoff S. 2001. *Lichens of North America.* Yale University Press.

Brundrett M C, Tedersoo T. 2018. Evolutionary history of mycorrhizal symbioses and global host plant diversity. *New Phytologist* 220: 1108–1115 (OA).

Burki F. 2014. The eukaryotic tree of life from a global phylogenomic perspective. *Cold Spring Harbor Perspectives in Biology* 6(5): a016147 (OA).

Burnett JH. 1976. *Fundamentals of Mycology*. 2nd ed. Edward Arnold.

Butt TM, Jackson C, Magan N (eds). 2001. *Fungi as Biocontrol Agents: Progress, Problems and Potential*. CABI.

CABI Invasive Species Compendium. n.d. *Hymenoscyphus fraxineus* (ash dieback). Accessed June 14, 2021. cabi.org/isc/datasheet/108083.

Cairns TC, Nai C, Meyer V. 2018. How a fungus shapes biotechnology: 100 years of *Aspergillus niger* research. *Fungal Biology and Biotechnology* 5: article 13 (OA).

Canadian Construction Association. 2018. *Mould Guidelines for the Canadian Construction Industry*. Available at cca-acc.com/wp-content/uploads/2019/02/Mould-guidelines2018.pdf.

Carbonetto B, Ramsayer J, Nidelet T, et al. 2018. Bakery yeasts, a new model for studies in ecology and evolution. *Yeast* 35: 591–603.

Casadevall A. 2012. Fungi and the rise of mammals. *PLOS Pathogens* 8(8): article e1002808 (OA).

Casselman A. 2007. Strange but true: The largest organism on Earth is a fungus. *Scientific American*, October 4, 2007. scientificamerican. com/article/strange-but-true-largest-organism-is-fungus.

Casselman KD. 2001. *Lichen Dyes: The New Source Book*. Courier Corp.

Centers for Disease Control and Prevention. 2020a. Antibiotic/anti-microbial resistance (AR/AMR). cdc.gov/drugresistance.

Centers for Disease Control and Prevention. 2020b. People living with HIV/AIDS. www.cdc.gov/fungal/infections/hiv-aids.html.

Cepelewicz J, Quanta Magazine. 2020. The case of the missing bacteria. *The Atlantic*, April 18, 2020. theatlantic.com/science/archive/2020/04/animals-microbiome-gut-bacteria/610201.

Cerimi K, Akkaya KC, Pohl C, et al. 2019. Fungi as source for new bio-based materials: A patent review. *Fungal Biology and Biotechnology* 6: article 17 (OA).

Chakrabarty P. 2018. Four billion years of evolution in six minutes. Filmed April 2018 in Vancouver, BC. TED2018 video, 5:32. ted.com/talks/prosanta_chakrabarty_four_billion_years_of_evolution_in_six_minutes.

Charles K. 2021. Frogs are battling their own terrible pandemic—can we stop it? *New Scientist*, July 14, 2021. newscientist.com/article/mg25133434-200-frogs-are-battling-their-own-terrible-pandemic-can-we-stop-it.

Chen J, Blume H-P, Beyer L. 2000. Weathering of rocks induced by lichen colonization—a review. *Catena* 39: 121–146.

Cheruiyot GK, Wanyonyi WC, Kiplimo JJ, et al. 2019. Adsorption of toxic crystal violet dye using coffee husks: Equilibrium, kinetics and thermodynamics study. *Scientific African* 5: article e00116 (OA).

Chial H. 2008. DNA sequencing technologies key to the Human Genome Project. *Nature Education* 1: 219.

Christian D. 2011. The history of our world in eighteen minutes. Filmed March 2011 in Long Beach, CA. TED2011 video, 17:24. ted.com/talks/david_christian_the_history_of_our_world_in_18_minutes.

Clay K. 1990. Fungal endophytes of grasses. *Annual Review of Ecology and Systematics* 21: 275-297 (OA).

Cordeiro RA (ed). 2019. *Pocket Guide to Mycological Diagnosis.* CRC Press.

Cui L, Morris A, Ghedin E. 2013. The human mycobiome in health and disease. *Genome Medicine* 5(7): article 63 (OA).

Czerucka D, Piche T, Rampal P. 2007. Yeast as probiotics—*Saccharomyces boulardii. Alimentary Pharmacology and Therapeutics* 26: 767-778 (OA).

D'Arcy CJ. 2000 (updated 2005). Dutch elm disease. *The Plant Health Instructor.* doi.org/10.1094/PHI-I-2000-0721-02 (OA).

de Oliveira Junqueira AC, de Melo Pereira GV, Coral Medina JD, et al. 2019. First description of bacterial and fungal communities in Colombian coffee beans fermentation analysed using Illumina-based amplicon sequencing. *Scientific Reports* 9: article 8794 (OA).

Desroches TC, McMullin DR, Miller JD. 2014. Extrolites of *Wallemia sebi,* a very common fungus in the built environment. *Indoor Air* 24: 533-542.

de Waal F. 2019. *Mama's Last Hug: Animal Emotions and What They Tell Us About Ourselves.* WW Norton & Co.

Diamond J. 1997. *Guns, Germs, and Steel.* WW Norton & Co.

Dijksterhuis J, Samson RA (eds). 2007. *Food Mycology: A Multifaceted Approach to Fungi and Food.* CRC Press.

Duan S-F, Han P-J, Wang Q-M, et al. 2018. The origin and adaptive evolution of domesticated populations of yeast from Far East Asia. *Nature Communications* 9: article 2690 (OA).

Dunn R. 2018. *Never Home Alone: From Microbes to Millipedes, Camel Crickets, and Honeybees, the Natural History of Where We Live.* Basic Books.

Dykes J. 2021. Mycoremediation: The under-utilised art of fungi clean-ups. *Geographical,* February 26, 2021. geographical.co.uk/

nature/climate/item/3980-mycoremediation-using-mushrooms-
to-clean-up-after-humans-could-be-an-under-utilised-opportunity.

Ekblad A, Wallander H, Godbold DL, et al. 2013. The production and
turnover of extramatrical mycelium of ectomycorrhizal fungi in
forest soils: Role in carbon cycling. *Plant and Soil* 366: 1–27.

Enaud R, Vandenborght L-E, Coron N, et al. 2018. The mycobiome:
A neglected component in the microbiota-gut-brain axis. *Microorganisms* 6(1): article 22 (OA).

EPA (United States Environmental Protection Agency). 2008. *Mold
Remediation in Schools and Commercial Buildings*. Available at
www.epa.gov/mold/mold-remediation-schools-and-commercial-
buildings-guide-chapter-1.

EPA (United States Environmental Protection Agency). 2021. Indoor
air quality (IAQ): What is a HEPA filter? Last updated March 3, 2021.
epa.gov/indoor-air-quality-iaq/what-hepa-filter-1.

Eschliman D, Ettlinger S. 2015. *Ingredients: A Visual Exploration of 75
Additives and 25 Food Products*. Regan Arts.

Ewbank A. 2018. Inside the world's only sourdough library. Atlas
Obscura, May 16, 2018. atlasobscura.com/articles/sourdough-
library.

Fiorilli V, Vannini C, Ortolani F., et al. 2018. Omics approaches revealed
how arbuscular mycorrhizal symbiosis enhances yield and resistance
to leaf pathogen in wheat. *Scientific Reports* 8: article 9625 (OA).

Fisher MC, Gow NAR, Gurr SJ. 2016. Tackling emerging fungal threats
to animal health, food security and ecosystem resilience. *Philosophical Transactions of the Royal Society B: Biological Sciences* 371: article
20160332 (OA).

Flad V, Young D, Seppälä S, et al. 2020. The biotechnological potential
of anaerobic gut fungi. In Benz JP, Schipper K (eds), *The Mycota*,
413–437. Vol. 2 of *Genetics and Biotechnology*. 3rd ed. Springer.

Flannigan B, Samson RA, Miller JD (eds). 2003. *Microorganisms in
Home and Indoor Work Environments: Diversity, Health Impacts,
Investigation and Control*. CRC Press.

Fleming A. 1929. On the antibacterial action of cultures of a *Penicillium*,
with special reference to their use in the isolation of *B. influenzae*.
British Journal of Experimental Pathology 10: 226–236 (OA).

Fortey R. 2008. *Dry Storeroom No. 1: The Secret Life of the Natural
History Museum*. Harper Perennial.

Franc GD. 2007. Potato wart. APS*net* Features. apsnet.org/edcenter/apsnetfeatures/Pages/PotatoWart.aspx.

Fraser JA, Heitman J. 2003. Fungal mating-type loci. *Current Biology* 13: R792–R795 (OA).

Garaudée S, Elhabiri M, Kalny D, et al. 2002. Allosteric effects in norbadione A: A clue for the accumulation process of ^{137}Cs in mushrooms? *Chemical Communications* 9: 944–945.

García-Estrada C, Cat E, Santamarta I. 2016. *Beauveria bassiana* as biocontrol agent: Formulation and commercialization for pest management. In Singh HB, Sarma BK, Keswani C (eds), *Agriculturally Important Microorganisms: Commercialization and Regulatory Requirements in Asia*, 81–96. Springer.

Geils BW, Hummer KE, Hunt RS. 2010. White pines, *Ribes*, and blister rust: A review and synthesis. *Forest Pathology* 40: 147–185 (OA).

Gepts P. 2004. Who owns biodiversity, and how should the owners be compensated? *Plant Physiology* 134: 1295–1307 (OA).

Gevers D, Knight R, Petrosino JF, et al. 2012. The Human Microbiome Project: A community resource for the healthy human microbiome. *PLOS Biology* 10: article e1001377 (OA).

Ghabrial SA, Castón JR, Jiang D, et al. 2015. 50-plus years of fungal viruses. *Virology* 479–480: 356–368 (OA).

Gibbons JG. 2019. How to train your fungus. *mBio* 10(6): article e03031-19 (OA).

Gibbons JG, Salichos L, Slot JC, et al. 2012. The evolutionary imprint of domestication on genome variation and function of the filamentous fungus *Aspergillus oryzae. Current Biology* 22: 1403–1409 (OA).

Grant EHC, Muths E, Katz RA, et al. 2016. *Salamander Chytrid Fungus (Batrachochytrium salamandrivorans) in the United States—Developing Research, Monitoring, and Management Strategies.* U.S. Geological Survey Open-File Report 2015-1233. doi.org/10.3133/ofr20151233.

Gräser Y, Monod M, Bouchara JP, et al. 2018. New insights in dermatophyte research. *Medical Mycology* 56(suppl. 1): S2–S9.

Griffin DW, Kellogg CA, Shinn EA. 2001. Dust in the wind: Long range transport of dust in the atmosphere and its implications for global public and ecosystem health. *Global Change and Human Health* 2: 20–33.

Hale MD, Eaton RA. 1985. Oscillatory growth of fungal hyphae in wood cell walls. *Transactions of the British Mycological Society* 84: 277–288.

Hall C, Welch J, Kowbel DJ, et al. 2010. Evolution and diversity of a fungal self/nonself recognition locus. *PLOS One* 5(11): article e14055 (OA).

Hall RA, Noverr MC. 2017. Fungal interactions with the human host: Exploring the spectrum of symbiosis. *Current Opinion in Microbiology* 40: 58–64 (OA).

Hammerstein P (ed). 2003. *Genetic and Cultural Evolution of Cooperation.* MIT Press.

Harari YN. 2014. *Sapiens: A Brief History of Humankind.* Harper Perennial.

Hariharan J. 2021. Guardians of the Microbial Galaxy. *Scientific American*, March 28, 2021. scientificamerican.com/article/guardians-of-the-microbial-galaxy.

Hawksworth DL, Rose F. 1976. *Lichens as Pollution Monitors.* Institute of Biology, Studies in Biology no. 66. Edward Arnold.

Hay WD. 1887. *An Elementary Text-Book of British Fungi.* Swan, Sonnenschein, Lowrey & Co. biodiversitylibrary.org/bibliography/4073.

Heusler K, Pletscher A. 2001. The controversial early history of cyclosporin. *Swiss Medical Weekly* 131(21–22): 299–302 (OA).

Holden EM. 2003. *Recommended English Names for Fungi in the UK.* Report to the British Mycological Society, English Nature, Plantlife and Scottish Natural Heritage. plantlife.org.uk/uk/our-work/publications/recommended-english-names-fungi-uk.

Hölldobler B, Wilson EO. 2010. *The Leafcutter Ants: Civilization by Instinct.* WW Norton & Co.

Holmes FW, Heybroek HM (trans). 1990. *Dutch Elm Disease—the Early Papers: Selected Works of Seven Dutch Women Phytopathologists.* American Phytopathological Society Press.

Holmes H. 2003. *The Secret Life of Dust: From the Cosmos to the Kitchen Counter, the Big Consequences of Little Things.* Wiley.

Holt S, Mukherjee V, Lievens B, et al. 2018. Bioflavoring by nonconventional yeasts in sequential beer fermentations. *Food Microbiology* 72: 55–66 (OA).

Horner EW, Miller JD. 2003. Microbial volatile organic compounds with emphasis on those arising from filamentous fungal contaminants of buildings. *ASHRAE Transactions* 109: 215–231.

Houbraken J, Frisvad JC, Samson RA. 2011. Fleming's penicillin producing strain is not *Penicillium chrysogenum* but *P. rubens.* *IMA Fungus* 2: 87–95 (OA).

Hoyt E. 1996. *The Earth Dwellers: Adventure in the Land of Ants.* Simon & Schuster.

Irmer J. 2017. Lignin—a natural resource with huge potential. Bioeconomy BW. biooekonomie-bw.de/en/articles/dossiers/lignin-a-natural-resource-with-huge-potential.

Jacobs R. 2019. *The Truffle Underground: A Tale of Mystery, Mayhem, and Manipulation in the Shadowy Market of the World's Most Expensive Fungus.* Clarkson Potter.

Jain U, Ver Heul AM, Xiong S, et al. 2021. *Debaryomyces* is enriched in Crohn's disease intestinal tissue and impairs healing in mice. *Science* 371: 1154–1159.

Kauffman CA. 2007. Histoplasmosis: A clinical and laboratory update. *Clinical Microbiology Reviews* 20: 115–132 (OA).

Keeling PJ, Burki F. 2019. Progress towards the tree of eukaryotes. *Current Biology* 29(16): R808–R817 (OA).

Kendrick B. 2017. *The Fifth Kingdom: An Introduction to Mycology.* 4th ed. Hackett Publishing.

Kennedy G. 1956. The two Explorer stratosphere balloon flights. StratoCat. stratocat.com.ar/artics/explorer-e.htm.

Kirkland TN, Fierer J. 2018. *Coccidioides immitis* and *posadasii*: A review of their biology, genomics, pathogenesis, and host immunity. *Virulence* 9: 1426–1435 (OA).

Koide RT, Mosse B. 2004. A history of research on arbuscular mycorrhiza. *Mycorrhiza* 14: 145–163.

Kokkoris V, Stefani F, Dalpé Y, et al. 2020. Nuclear dynamics in the arbuscular mycorrhizal fungi. *Trends in Plant Science* 25: 765–778.

Kolbert E. 2014. *The Sixth Extinction: An Unnatural History.* Picador Books.

Kumar P, Mahato DK, Kamle M, et al. 2017. Aflatoxins: A global concern for food safety, human health and their management. *Frontiers in Microbiology* 7: article 2170 (OA).

Lange L. 2017. Fungal enzymes and yeasts for conversion of plant biomass to bioenergy and high-value products. In Heitman J, Howlett BJ, Crous PW, et al. (eds), *The Fungal Kingdom*, 1029–1048. ASM Press.

Laniado-Laborín R, Cabrales-Vargas MN. 2009. Amphotericin B: Side effects and toxicity. *Revista Iberoamericana de Micología* 26: 223–227 (OA).

Large EC. 1940. *The Advance of the Fungi.* Jonathan Cape.

Lax E. 2004. *The Mold in Dr. Florey's Coat: The Story of the Penicillin Miracle*. Henry Holt.

Leggett M, Cross J, Hnatowich G, et al. 2007. Challenges in commercializing a phosphate-solubilizing microorganism: *Penicillium bilaiae*, a case history. In Velázquez E, Rodríguez-Barrueco C (eds), *First International Meeting on Microbial Phosphate Solubilization*, 215–222. Springer.

Lichtwardt RW. 2012. *The Trichomycetes: Fungal Associates of Arthropods*. Springer-Verlag.

Limper AH, Adenis A, Le T, et al. 2017. Fungal infections in HIV/AIDS. *The Lancet Infectious Diseases* 17 (11): e334–e343 (OA).

Lincoff G, Mitchel DH. 1977. *Toxic and Hallucinogenic Mushroom Poisoning: A Handbook for Physicians and Mushroom Hunters*. Van Nostrand Reinhold.

Lindeberg G. 1989. Elias Melin: Pioneer leader in mycorrhizal research. *Annual Review of Phytopathology* 27: 49–58 (OA).

Lipka S. 2009. When corn tastes like mushrooms. *The Atlantic*, October 26, 2009. theatlantic.com/health/archive/2009/10/when-corn-tastes-like-mushrooms/28941.

Lomer CJ, Bateman RP, Johnson DL, et al. 2001. Biological control of locusts and grasshoppers. *Annual Review of Entomology* 46: 667–702.

Ludlow CL, Cromie GA, Garmendia-Torres C, et al. 2016. Independent origins of yeast associated with coffee and cacao fermentation. *Current Biology* 26: 965–971 (OA).

Maheswaran D, Zeng Y, Chan-Yeung M, et al. 2014. Exposure to beta-(1,3)-D-glucan in house dust at age 7–10 is associated with airway hyperresponsiveness and atopic asthma by age 11–14. *PLOS One* 9 (6): article e98878 (OA).

Malone D. 2018. Fungus may be key to colonizing Mars. *Building Design & Construction*, July 13, 2018. bdcnetwork.com/fungus-may-be-key-colonizing-mars.

Mann CC. 2018. *The Wizard and the Prophet: Two Remarkable Scientists and Their Dueling Visions to Shape Tomorrow's World*. Knopf.

Marasas WFO. 1995. Fumonisins: Their implications for human and animal health. *Natural Toxins* 3: 193–198.

Margulis L. 1998. *Symbiotic Planet: A New Look at Evolution*. Basic Books.

Margulis L, Sagan D. 1986. *Microcosmos: Four Billion Years of Microbial Evolution*. University of California Press.

Martin W, Mentel M. 2010. The origin of mitochondria. *Nature Education* 3(9): 58 (OA).

Matossian MK. 1982. Ergot and the Salem witchcraft affair: An outbreak of a type of food poisoning known as convulsive ergotism may have led to the 1692 accusations of witchcraft. *American Scientist* 70: 355-357.

Matossian MK. 1991. *Poisons of the Past: Molds, Epidemics, and History*. Revised ed. Yale University Press.

MatterofTrust.org. n.d. Renewable resources – Fungi + SF oil spill hair mats bioremediation P.1 – 2007-2008. Accessed June 15, 2021. matteroftrust.org/oily-hair-mat-remediation-sf-bay-area-treat ability-study-phase-i-completed.

May RC, Stone NRH, Wiesner DL, et al. 2016. *Cryptococcus*: From environmental saprophyte to global pathogen. *Nature Reviews Microbiology* 14: 106-117.

McCallum BD, DePauw RM. 2008. A review of wheat cultivars grown in the Canadian prairies. *Canadian Journal of Plant Science* 88: 649-677 (OA).

McCoy P. 2016. *Radical Mycology: A Treatise on Seeing and Working With Fungi*. Chthaeus Press.

McCutchen M. 2017. The five most expensive truffles ever. Money Inc. moneyinc.com/five-expensive-truffles-ever.

McKenna M. 2020. Coffee rust is going to ruin your morning. *The Atlantic*, September 16, 2020. theatlantic.com/science/archive/ 2020/09/coffee-rust/616358.

Meis JF, Chowdhary A, Rhodes JL, et al. 2016. Clinical implications of globally emerging azole resistance in *Aspergillus fumigatus*. *Philosophical Transactions of the Royal Society B: Biological Sciences* 371: article 20150460 (OA).

Metcalfe T. 2016. Oldest beer brewed from shipwreck's 220-year-old yeast microbes. Live Science, November 10, 2016. livescience. com/56814-oldest-beer-recreated-from-shipwreck-yeast.html.

Meyer V, Basenko EY, Benz JP, et al. 2020. Growing a circular economy with fungal biotechnology: A white paper. *Fungal Biology and Biotechnology* 7: article 5 (OA).

Miller JD. 2019. The role of dust mites in allergy. *Clinical Reviews in Allergy and Immunology* 57: 312-329.

Miller JD, MacKenzie S. 2000. Secondary metabolites of *Fusarium venenatum* strains with deletions in the *Tri5* gene encoding trichodiene synthetase. *Mycologia* 92: 764–771.

Miller JD, Rand TG, Jarvis BB. 2003. *Stachybotrys chartarum*: Cause of human disease or media darling? *Medical Mycology* 41: 271–291.

Money NP. 2018. *The Rise of Yeast: How the Sugar Fungus Shaped Civilization*. Oxford University Press.

Moore D, Robson GD, Trinci APJ. 2020. *21st Century Guidebook to Fungi*. 2nd ed. Cambridge University Press.

Mustafa AM, Angeloni S, Nzekoue FK, et al. 2020. An overview on truffle aroma and main volatile compounds. *Molecules* 25(24): article 5948 (OA).

Nagel T. 1974. What is it like to be a bat? *The Philosophical Review* 83: 435–450.

NationMaster. 2019. Export of blue-veined cheese. nationmaster.com/nmx/ranking/export-of-blue-veined-cheese.

Nelsen MP, Lücking R, Boyce CK, et al. 2020. No support for the emergence of lichens prior to the evolution of vascular plants. *Geobiology* 18: 3–13.

Nielsen J. 2013. Production of biopharmaceutical proteins by yeast: Advances through metabolic engineering. *Bioengineered* 4: 207–211 (OA).

Nielsen KF, Sumarah MW, Frisvad JC, et al. 2006. Production of metabolites from the *Penicillium roqueforti* complex. *Journal of Agricultural and Food Chemistry* 54: 3756–3763.

Nygaard S, Hu H, Li C, et al. 2016. Reciprocal genomic evolution in the ant-fungus agricultural symbiosis. *Nature Communications* 7: article 12233 (OA).

OED Online. 2021. Mycophile. *Oxford English Dictionary*. Oxford University Press.

Oerke E-C. 2020. Remote sensing of diseases. *Annual Review of Phytopathology* 58: 225–252.

O'Hanlon SJ, Rieux A, Farrer RA, et al. 2018. Recent Asian origin of chytrid fungi causing global amphibian declines. *Science* 360: 621–627 (OA).

Oldest.org. n.d. Oldest professions in the world. Accessed June 14, 2021. oldest.org/people/professions.

Palmer WA, Heard TA, Sheppard AW. 2010. A review of Australian classical biological control of weeds programs and research activities over the past 12 years. *Biological Control* 52: 271–387.

Pataky J, Snetselaar K. 2006. Common smut of corn. *The Plant Health Instructor*. doi.org/10.1094/PHI-I-2006-0927-01 (OA).

Patrick J. 2013. The ant city. YouTube video posted August 25, 2013, 3:25. (Clip from the documentary *Ants: Nature's Secret Power*, dir. Thaler W, 2004.) youtube.com/watch?v=lKD4aXNOOC0.

Peter J, De Chiara M, Friedrich A, et al. 2018. Genome evolution across 1,011 *Saccharomyces cerevisiae* isolates. *Nature* 556: 339–344 (OA).

Petersen JH. 2013. *The Kingdom of Fungi*. Princeton University Press.

Pettersson OV, Leong SL. 2011. Fungal xerophiles (osmophiles). *eLS*. doi.org/10.1002/9780470015902.a0000376.pub2.

Pirttilä AM, Frank AC (eds). 2011. *Endophytes of Forest Trees: Biology and Applications*. Springer.

Pitt JI, Wild CP, Baan RA, et al. (eds). 2012. *Improving Public Health Through Mycotoxin Control*. International Agency for Research on Cancer, Scientific Publication no. 158. publications. iarc.fr/Book-And-Report-Series/Iarc-Scientific-Publications/Improving-Public-Health-Through-Mycotoxin-Control-2012.

Pollan M. 2018. *How to Change Your Mind: What the New Science of Psychedelics Teaches Us About Consciousness, Dying, Addiction, Depression, and Transcendence*. Penguin Books.

Porter JR. 1973. Agostino Bassi bicentennial (1773–1973). *Bacteriological Reviews* 37: 284–288 (OA).

Powers R. 2018. *The Overstory: A Novel*. WW Norton & Co.

Pretorius ZA, Singh RP, Wagoire WW, et al. 2000. Detection of virulence to wheat stem rust resistance gene *Sr31* in *Puccinia graminis* f. sp. *tritici* in Uganda. *Plant Disease* 84: 203 (OA).

Pringle P. 2013. *Experiment Eleven: Dark Secrets Behind the Discovery of a Wonder Drug*. Bloomsbury Press.

Rayner ADM. 1997. *Degrees of Freedom: Living in Dynamic Boundaries*. Imperial College Press.

Rayner ADM, Boddy L. 1988. *Fungal Decomposition of Wood: Its Biology and Ecology*. John Wiley & Sons.

Richardson DHS. 1975. *The Vanishing Lichens: Their History, Biology and Importance*. David & Charles.

Rogers A. 2015. *Proof: The Science of Booze*. Mariner Books.

Rokas A. 2018. Where sexes come by the thousands. The Conversation, October 30, 2018. theconversation.com/where-sexes-come-by-the-thousands-105554.

Rosenblum Lichtenstein JH, Hsu Y-H, Gavin IM, et al. 2015. Environmental mold and mycotoxin exposures elicit specific cytokine and chemokine responses. *PLOS One* 10(5): article e0126926 (OA).

Ryan F. 2002. *Darwin's Blind Spot: Evolution Beyond Natural Selection.* Houghton Mifflin.

Ryan FJ, Beadle GW, Tatum EL. 1943. The tube method of measuring the growth rate of *Neurospora. American Journal of Botany* 30: 784–799.

Salehi Jouzani G, Tabatabaei M, Aghbashlo M (eds). 2020. *Fungi in Fuel Biotechnology.* Springer.

Sánchez C, Moore D, Robson G, et al. 2020. A 21st century mini-guide to fungal biotechnology. *Mexican Journal of Biotechnology* 5: 11–42 (OA).

Saunders CW, Scheynius A, Heitman J. 2012. *Malassezia* fungi are specialized to live on skin and associated with dandruff, eczema, and other skin diseases. *PLOS Pathogens* 8(6): article e1002701 (OA).

Savage S. n.d. Infographic: 9 plant diseases that threaten your favorite foods—and how GM can help. Genetic Literacy Project. Accessed June 14, 2021. geneticliteracyproject.org/2014/08/12/infographic-9-plant-diseases-that-threaten-your-favorite-foods-and-how-gm-can-help.

Schardl CL, Phillips TD. 1997. Protective grass endophytes: Where are they from and where are they going? *Plant Disease* 81: 430–438 (OA).

Schilthuizen M. 2018. *Darwin Comes to Town: How the Urban Jungle Drives Evolution.* Picador.

Schrenk D, Bignami M, Bodin L, et al. (EFSA Panel on Contaminants in the Food Chain [CONTAM]). 2020. Risk assessment of aflatoxins in food. *EFSA Journal* 18(3): article e06040 (OA).

Schumann GL, Leonard KJ. 2011. Stem rust of wheat (black rust). *The Plant Health Instructor.* doi.org/10.1094/PHI-I-2000-0721-01 (OA).

Schumann GL, Uppala S. 2017. Ergot of rye. *The Plant Health Instructor.* apsnet.org/edcenter/disandpath/fungalasco/pdlessons/Pages/Ergot.aspx (OA).

Schwartz IS, Kauffman CA. 2020. Blastomycosis. *Seminars in Respiratory and Critical Care Medicine* 41: 31–41.

Schwartzberg L (dir). 2019. *Fantastic Fungi*. Moving Art Studio, Reconsider.

Scott PM, Kanhere SR. 1979. Instability of PR toxin in blue cheese. *Journal of Association of Official Analytical Chemists* 62: 141–147.

Seed PC. 2015. The human mycobiome. *Cold Spring Harbor Perspectives in Medicine* 5: article a019810 (OA).

Seifert KA. 2013. Memorials to the great famine. *IMA Fungus* 4(2): A50–A54 (OA).

Sender R, Fuchs S, Milo R. 2016. Are we really vastly outnumbered? Revisiting the ratio of bacterial to host cells in humans. *Cell* 164: 337–340 (OA).

Sheldrake M. 2020. *Entangled Life: How Fungi Make Our Worlds, Change Our Minds and Shape Our Futures*. Random House.

Shrestha G, St. Clair LL. 2013. Lichens: A promising source of antibiotic and anticancer drugs. *Phytochemistry Reviews* 12: 229–244.

Shunk GK, Gomez XR, Averesch NJH. 2020 (preprint). A self-replicating radiation-shield for human deep-space exploration: Radiotrophic fungi can attenuate ionizing radiation aboard the International Space Station. *bioRxiv* 2020.07.16.205534 (OA).

Shurtleff W, Aoyagi A. 1979. *The Book of Tempeh*. Soyinfo Center.

Shurtleff W, Aoyagi A. 2012. *History of Koji—Grains and/or Soybeans Enrobed With a Mold Culture (300 BCE to 2012): Extensively Annotated Bibliography and Sourcebook*. Soyinfo Center. Available at soyinfocenter.com/pdf/154/Koji.pdf.

Sibley DA. 2020. *What It's Like to Be a Bird: From Flying to Nesting, Eating to Singing—What Birds Are Doing, and Why*. Knopf.

Simard SW. 2018. Mycorrhizal networks facilitate tree communication, learning, and memory. In Baluška F, Gagliano M, Witzany G (eds), *Memory and Learning in Plants*, 191–213. Springer.

Simard S. 2021. *Finding the Mother Tree: Discovering the Wisdom of the Forest*. Allen Lane.

Simard SW, Perry DA, Jones MD, et al. 1997. Net transfer of carbon between ectomycorrhizal tree species in the field. *Nature* 388: 579–582 (OA).

Smith AH. 1973. *The Mushroom Hunter's Field Guide*. University of Michigan Press.

Smith DM. 1877. *Arctic Expeditions From British and Foreign Shores: From the Earliest Times to the Expedition of 1875–76*. Thomas C. Jack, Grange Publishing Works.

Smith SE, Read DJ. 2008. *Mycorrhizal Symbiosis*. 3rd ed. Academic Press.

Sobrova P, Adam V, Vasatkova A, et al. 2010. Deoxynivalenol and its toxicity. *Interdisciplinary Toxicology* 3: 94–99 (OA).

Sokulska M, Kicia M, Wesołowska M, et al. 2015. *Pneumocystis jirovecii*—from a commensal to pathogen: Clinical and diagnostic review. *Parasitology Research* 114: 3577–3585 (OA).

Stamets P. 2005. *Mycelium Running: How Mushrooms Can Help Save the World*. Ten Speed Press.

Stamets P. 2008. 6 ways mushrooms can save the world. Filmed March 2008 in Monterey, California. TED2008 video, 17:25. ted.com/talks/paul_stamets_6_ways_mushrooms_can_save_the_world.

Stamets P (ed). 2019. *Fantastic Fungi: Expanding Consciousness, Alternative Healing, Environmental Impact*. Earth Aware Editions.

Stamets P, Chilton JS. 1983. *The Mushroom Cultivator: A Practical Guide to Growing Mushrooms at Home*. Agarikon Press.

Stringer JR, Beard CB, Miller RF, et al. 2002. A new name (*Pneumocystis jiroveci*) for *Pneumocystis* from humans. *Emerging Infectious Diseases* 8: 891–896 (OA).

Suttle CA. 2013. Viruses: Unlocking the greatest biodiversity on Earth. *Genome* 56: 542–544 (OA).

Tanney JB, McMullin DR, Miller JD. 2018. Toxigenic foliar endophytes from the Acadian forest. In Pirttilä AM, Frank AC (eds), *Endophytes of Forest Trees: Biology and Applications*, 343–381. Springer.

Tischer C, Chen C-M, Heinrich J. 2011. Association between domestic mould and mould components, and asthma and allergy in children: A systematic review. *European Respiratory Journal* 38: 812–824 (OA).

Tkavc R, Matrosova VY, Grichenko OE, et al. 2018. Prospects for fungal bioremediation of acidic radioactive waste sites: Characterization and genome sequence of *Rhodotorula taiwanensis* MD1149. *Frontiers in Microbiology* 8: article 2528 (OA).

Trappe JM. 2005. A.B. Frank and mycorrhizae: The challenge to evolutionary and ecologic theory. *Mycorrhiza* 15: 277–281.

Tsing AL. 2015. *The Mushroom at the End of the World: On the Possibility of Life in Capitalist Ruins*. Princeton University Press.

UNAIDS. 2021. Fact sheet 2021: Global HIV statistics. unaids.org/sites/default/files/media_asset/UNAIDS_FactSheet_en.pdf.

United Nations Environment Programme. 2016. Invasive species—a huge threat to human well-being. unep.org/news-and-stories/story/invasive-species-huge-threat-human-well-being.

Van Asselt L. 1999. Review: Interactions between domestic mites and fungi. *Indoor and Built Environment* 8: 216–220.

VanderGoot J. 2017. Considering dry rot: The co-evolution of buildings and *Serpula lacrymans*. *Journal of Architectural Education* 71: 225–231.

Verweij PE, Snelders E, Kema GH, et al. 2009. Azole resistance in *Aspergillus fumigatus*: A side-effect of environmental fungicide use? *The Lancet Infectious Diseases* 9: 789–795.

Wagner SC. 2011. Biological nitrogen fixation. *Nature Education Knowledge* 3: article 15 (OA).

Ward HM. 1888. Anton De Bary. *Nature* 37: 297–299 (OA).

Wasson RG. 1972. The death of Claudius or mushrooms for murderers. *Botanical Museum Leaflets*, Harvard University, 23: 101–128. Available at biodiversitylibrary.org/part/168556.

Wasson VP, Wasson RG. 1957. *Mushrooms, Russia, and History*. Pantheon.

Webster J, Weber R. 2007. *Introduction to Fungi*. 3rd ed. Cambridge University Press.

Wiebe MG. 2004. Quorn™ myco-protein—Overview of a successful fungal product. *Mycologist* 18: 17–20.

Wipf D, Krajinski F, van Tuinen D, et al. 2019. Trading on the arbuscular mycorrhiza market: From arbuscules to common mycorrhizal networks. *New Phytologist* 223: 1127–1142 (OA).

Wohlleben P. 2016. *The Hidden Life of Trees: What They Feel, How They Communicate—Discoveries From a Secret World*. Greystone Books.

World Health Organization. 2018. *Fumonisins*. Food Safety Digest. Department of Food Safety and Zoonoses. who.int/foodsafety/FSDigest_Fumonisins_EN.pdf.

World Health Organization. n.d. Antimicrobial resistance. who.int/health-topics/antimicrobial-resistance.

Wright C, Gryganskyi AP, Bonito G. 2016. Fungi in composting. In Purchase D (ed), *Fungal Applications in Sustainable Environmental Biotechnology*, 3–28. Springer.

Yin X, Yang A-A, Gao J-M. 2019. Mushroom toxins: Chemistry and toxicology. *Journal of Agricultural and Food Chemistry* 67: 5053–5071.

Yong E. 2016. *I Contain Multitudes: The Microbes Within Us and a Grander View of Life.* Ecco.

Yong E. 2017. How the zombie fungus takes over ants' bodies to control their minds. *The Atlantic,* November 14, 2017. theatlantic.com/ science/archive/2017/11/how-the-zombie-fungus-takes-over-ants-bodies-to-control-their-minds/545864.

Zabel RA, Morrell JJ. 2012. *Wood Microbiology: Decay and Its Prevention.* Academic Press.

Zhang S. 2017. The secrets of the "humongous fungus": How one of the biggest living organisms in the world got so big. *The Atlantic,* October 30, 2017. theatlantic.com/science/archive/2017/10/ humongous-fungus-genome/544265.

Zupančič J, Babič MN, Zalar P, et al. 2016. The black yeast *Exophiala dermatitidis* and other selected opportunistic human fungal pathogens spread from dishwashers to kitchens. *PLOS One* 11(2): article e0148166 (OA).

INDEX

epiphytes, 60
ergosterol, 172
ergot, 95, 101–3
ergotamine, 102
ericoid mycorrhizae, 66
Escherichia coli (*E. coli*), 187
Escovopsis species, 43, 44, 227
ester, 117–18
ethanol. *See* alcohol
Eucalyptus (gum trees), 58,
    73, 171
eukaryotes, 14, 52
Eurotiales, 225–26
Eurotiomycetes, 225
evolution, 13–14, 15–16, 36
Exophiala species, 144–45, 225

F
farming. *See* agriculture
fermentation, 112–38; about, 113;
    alcohol, 112–13, 114–15, 116–18,
    194; bread, 120–21; cheese,
    121–26, 194–95, 245n21; choco-
    late, 129–30; coffee, 131; kōji
    mould, 8, 127–28, 129; kombu-
    cha, 119, 197–98; Pu-erh tea,
    130; secondary fermentation,
    119; sourdough, 121; soy sauce,
    126–27, 128, 241nn12–13; tem-
    peh, 137–38, 242n26
fescue foot, 95
flask fungi, 227–28
Fleming, Alexander, 181–82, 183,
    244n5
Florey, Harold, 182, 183
Florida perforate cladonia
    (*Cladonia perforata*), 217, 226
food: about, 111; compost and,
    134–36, 138; fungal protein,
    119, 187–88; future of, 197–98;
    mouldy, 131–33; mushrooms, 6,
    20, 136; preservation methods,
    132–34. *See also* fermentation

Food and Agriculture Organization
    (FAO), 103, 132
forests, 57–86; biocontrol for
    spruce budworm, 83–86; bio-
    degradation of forest wastes,
    188–89; carbon cycle, 79–80;
    commercial plantations, 57–58,
    73; decomposition, 58, 80–83;
    endophytes within trees, 59–64;
    evidence of fungal activity, 201;
    mycorrhizal networks among
    roots, 64–70; nurseries, 85; old-
    growth, 57, 73; reintroducing
    fungi into, 211; tree diseases,
    70–72, 73, 74–79; Trillion Tree
    Campaign, 215
Frank, Albert Bernhard, 65
frogs, 207–8
fumonisins, 106
fungal infection–mammalian selec-
    tion (FIMS) hypothesis, 158
fungaria, 209. *See also* biological
    collections
fungi: approach to, 6, 10, 218; Asco-
    mycota (ascomycetes), 16–17,
    19, 37, 80, 224–28; author's
    background with, 2–5, 19–20;
    Basidiomycota (basidiomycetes),
    16, 17–18, 19, 31, 80, 229–31;
    biggest, 71–72, 73; Chytridio-
    mycota (chytrids), 14–15, 19,
    223; citizen science and, 213–15;
    classification (taxonomy), 24,
    221–22, 234nn2–3; collections
    of, 209–10, 211–12, 238n22;
    communication among, 32–33,
    67–68, 237n7; cultural attitudes,
    8; DNA sequencing, 23–24
    (*see also* DNA); in dust, 1–2, 146;
    eating and digestion by, 29;
    evidence of, 201–2; evolution of,
    13–14, 15–16; growth, 28–29, 30,
    236n12; hyphae, 6, 13, 16, 25–26,

J

Janthinobacterium lividum (Jliv),
207-8
jelly fungi, 18, 65, 230
jock itch, 161

K

kingdoms, 14, 233n3, 234n2
Kloeckera species, 130, 225
Kluyveromyces species, 130, 225
kōji mould (*Aspergillus oryzae*),
8, 127-28, 129, 180, 188, 225,
241n13
kombucha, 119, 197-98

L

Lactarius species (milk caps),
66, 231
lactic acid, 131
lactic acid bacteria (*Lactobacillus*
species), 118, 119, 121, 123, 125,
130, 131
laundry detergent, 180-81
leaf-cutting ants, 41-44
Lecanoromycetes, 226. *See also*
lichens
Leotiomycetes (inoperculate
discomycetes), 226-27
Leptographium longiclavatum, 77,
228
Leucocoprinus species: *L. birnbaumii*, 231, 237n10; *L. gongylophorus*, 42, 231
leukoencephalomalacia (mouldy
corn disease), 106. See also
under *Fusarium* species:
F. verticillioides
lichens, 36-40; about, 6, 16; citizen
science and, 38; classification,
226; composition and structure, 37; dyes from, 39-40;
longevity and survival, 38-39;
medicinal uses, 40; mutualistic

relationships, 37-38; resources
on, 236n4; secondary metabolites from, 39; in Sudbury, 36-37
lighting, bioluminescent, 199
lignin, 17, 80, 81, 188-89
ligninase, 80, 81, 178, 189
lipase, 180
Listeria, 123
litmus lichen (*Roccella tinctoria*),
39, 226
loculoascomycetes, 227
locusts, 109
Lophodermium species, 57, 59, 61,
62, 84, 226
LSD (lysergic acid diethylamide),
102-3
lungs, 166-69
Lycoperdon species (puffballs), 229

M

Magnaporthales, 228
Malassezia species (dandruff), 7,
156, 160-61, 163, 197, 199, 229
Malasseziomycetes, 229
mantles, 57, 64-65
Marmite, 119
Mars, 196-97
materials, from mycelium, 195-97,
198-99
matsutake (*Tricholoma* species),
68-69, 191, 231, 238n10
medicine. *See* drugs
Melin, Elias, 67-68
metabolism, 32
metabolites, 32-33, 39, 40, 178,
184. *See also* microbial volatile
organic compounds; mycotoxins
Metarhizium acridum (green muscardine mould, Green Muscle),
109, 228
microbial volatile organic compounds (MVOCs), 139-40,
242n2